Herfried Griengl (ed.)

Biocatalysis

Springer-Verlag Wien GmbH

Prof. Dr. Herfried Griengl
Institute of Organic Chemistry, Technical University of Graz, Austria

Originally published by Springer-Verlag Wien New York in 2000

Typesetting: Thomson Press (India) Ltd., New Delhi

Printed on acid-free and chlorine-free bleached paper

With 51 Figures

CIP data applied for

Special Edition of
"Monatshefte für Chemie/Chemical Monthly", Vol. 131, No. 6, 2000

ISBN *978-3-211-83527-2* ISBN 978-3-7091-6310-8 (eBook)
DOI 10.1007/978-3-7091-6310-8

Editorial

Biocatalysis – the application of Nature's catalysts, the enzymes, to unnatural substrates – has become an important methodology in synthetic organic chemistry. Besides homo- and heterogenous catalysis, biocatalysis constitutes the third important group of catalytic methods available today. This technique enables extremely high selectivities to be achieved: excellent chemo-, regio-, enantio-, and stereoselectivities could be shown to be obtainable by using biocatalytic methods. In many cases, biocatalysis has proven to be the only approach to solve a given synthetic task. In addition, the use of enzymes, either isolated or as whole microorganism or plant cells, offer tremendous potential for ecologically clean production on an industrial scale.

For these reasons it is not surprising that, with respect to research and development, activity in this field is high on a world-wide scale. To give an overview on recent trends in biocatalysis, researchers in this active area have been invited to contribute.

The finding that enzymes are not only active in water but also in organic solvents was one of the great breakthroughs in biocatalysis which opened a wealth of preparative possibilities. Synthetically, approaches employing these enzymatic transformations are well developed. This is in contrast to the level of understanding the influence of the type of solvent and the content of water on the performance of a given enzyme. In this respect, the review from the group in Milan (*Giacomo Carrea* and coworkers) on the modelling of enzyme properties in organic solvents is of particular importance.

Hydrolases are still the most frequently used enzymes for biocatalytic transformations. From the Delft group (*Fred van Rantwijk* and colleagues), recent examples of developments in the field of lipase catalyzed synthesis of amides are reviewed.

Two other invited reviews are devoted to the application of biocatalysis to organic synthesis: the article on chemoenzymatic synthesis of lipidated peptides by *Herbert Waldmann* and *Dieter Kadereit* from Karlsruhe and a contribution from the Oviedo group (*Vicente Gotor* and *Miguel Ferrero*) on the use of enzymes for synthetic work in the area of nucleoside chemistry.

Regarding the original scientific contributions, the fact that hydrolases are the most commonly applied enzymes in biocatalysis is reflected in the number of articles contained in this issue demonstrating the current trends in biocatalysis. The question of regio- and stereoselectivity in enzyme-catalyzed reactions is considered in an article from another group from Milan (*Enzo Santaniello* and coworkers) dealing with the lipase catalyzed regio- and stereoselective acylation of hydroxy groups in steroidal side chains. The differentiation between substrates as well as the stereospecificity of transformations employing Penicillin G amidases is described by *Volker Kasche* and coworkers from Hamburg.

The preparation of enantiopure compounds is of tremendous importance in the area of chiral drugs and agrochemicals as well as in the flavour and fragrance fields. One approach still pursued is resolution of racemates employing hydrolytic enzymes. A representative example of this method is presented in the contribution by *Uwe T. Bornscheuer* and coworkers (Greifswald) concerning the resolution of the important drug ibuprofen using lipases. Another example is the article by *Kurt Faber* and coworkers (Graz) on the resolution of linalyl acetate. Here the question is raised how the difficulty of resolving tertiary alcohols might be overcome by novel esterase preparations. A further contribution by *Markus Pietzsch* and colleagues (Stuttgart, Zürich) is devoted to the deracemization of hydantoins.

Hydrolysis of nitriles is a biotransformation of tremendous industrial importance as the basis for the technical production of acrylamide. The question of manufacturing acrylic acid by using an immobilized biocatalyst is raised and discussed by *John Colby* and coworkers (Sunderland).

The introduction of hydroxyl groups onto unactivated carbon atoms in a regio- and enantioselective manner is only possible by using biohydroxylation. One of the most frequently used microorganisms for this purpose is *Beauveria bassiana* ATCC 7159. In the contribution from *Herbert Holland* (Canada) it is shown that also a *L*-specific α-amino acid benzamidase is present in this microorganism.

The microbial hydroxylation of aromatic compounds leading to cyclohexadiene-1,2-diols is one of the great achievements in biocatalysis. In their contribution, *Derek Boyd* and coworkers (Belfast) investigate the question of how this methodology can also be applied to acyclic dienes.

Another area of recent progress in biocatalysis is the use of peroxidases. Important contributions to this field are the articles from Amsterdam (*Ron Wever* and colleagues) and Würzburg (*Waldemar Adam* and coworkers). It is shown how the outcome of biotransformations can be influenced by crucial choice of enzyme and conditions.

It is hoped that this book will give a good impression on recent trends in this area. I would like to thank all colleagues from the field of biocatalysis for their contributions. The help of Profs. *Falk* and *Kalchhauser* is gratefully acknowledged.

Herfried Griengl

Contents

Invited Review

Modelling of Enzyme Properties in Organic Solvents

Giorgio Colombo, Gianluca Ottolina, and **Giacomo Carrea***

Istituto di Biocatalisi e Riconoscimento Molecolare CNR, I-20131 Milano, Italy

Summary. In this article we review how molecular modeling techniques can be used to shed some light on the influence of organic solvents on the molecular characteristics of proteins and enzymes. Molecular dynamic simulations on bovine pancreas trypsin inhibitor, chymotrypsin, and subtilisin make it possible to get a deeper understanding into how increased intramolecular interactions improve conformational rigidity, thus explaining the lower reactivity and the higher thermostability of enzymes in non-aqueous media. The application of thermodynamics-based models allows first qualitative predictions on the selectivity of many reaction types; however, the application of quantum mechanical/molecular mechanical methods is required for the development of quantitative models of actual reactivity patterns.

Keywords. Molecular modelling; Structure; Reactivity of enzymes; Organic solvents.

Introduction

The suspension or dissolution of enzymes in non-aqueous media affords many advantages (as well as some disadvantages) over the use of proteins in aqueous solution. For example, *Klibanov* and coworkers have demonstrated that proteins exhibit increased thermostability [1], molecular *pH* memory [2], and altered substrate specificity [3] when placed in an anhydrous organic solvent. Furthermore, the ability of enzymes to catalyze reactions that are either kinetically or thermodynamically impossible in water has also been observed [4–6]. This led to widespread use of enzymes as highly specific catalysts in non-aqueous solvents in the synthesis of organic compounds. For instance, a number of peptides containing *D*-amino acid residues have been synthesized using subtilisin as a catalyst, which is impossible in water because of the enzyme's strict *L*-stereoselectivity [7]. The synthetic potential of this phenomenon was also exploited in the subtilisin-catalyzed acylation of carbohydrates in anhydrous dimethylformamide. Not only does subtilisin esterify a number of sugars and related compounds in this solvent, but it does so on a preparative scale and with a marked positional selectivity [7]. Lipases and other hydrolases in non-aqueous

* Corresponding author

solvents also catalyze a variety of reactions like esterifications, transesterifications, aminolysis, and thio-transesterifications, whereas in water these processes are almost completely suppressed by hydrolysis [8].

The tyrosinase-catalyzed oxidation of phenols is another example. In water, this reaction proceeds with negligible yields due to the rapid polymerization of the resulting *ortho*-quinones and inactivation of the enzyme. In contrast, in chloroform, where both species are more stable, the quinones are readily obtained [9].

The possibility to apply biocatalysis to the synthesis of a wide range of compounds with high enantiomeric excesses as well as the new thermostability characteristics that enzymes can develop in organic solvents spurred investigation towards a deeper understanding of how protein structure and reactivity are affected by anhydrous organic solvents. The rationalization of these effects is supposed to enhance our ability to employ proteins as catalytic agents in non-aqueous media. This review will focus on different modelling approaches to the study of several properties of enzymes in these media. First, we will concentrate on enzyme structural properties in organic solvents, whereas in the second part the stress will be put on reactivity issues.

Protein Conformation in Organic Media

Flexibility and solvent accessible surface area

In addition to retaining catalytic activity, enzymes exhibit profound enhanced thermostability in organic solvents [10, 11]. Furthermore, it was found that in organic solutions it is possible to "imprint" the protein with an inhibitor and to induce changes in the enzymatic rate and selectivity after removal of the inhibitor [12, 13]. These characteristics of enzymes in organic solutions have been explained on the basis of increased rigidity of proteins in non-aqueous media. *Klibanov* and others have proposed that water acts as a "molecular lubricant", resulting in greater protein flexibility [14–17]. However, scant information is available at the molecular level to prove or to disprove this proposal. In order to accurately assess the role of flexibility in the activity of proteins, information concerning protein flexibility in both aqueous and nonaqueous environments was required [18].

Computer simulations have proven to be a valuable tool in the understanding of protein structure and dynamics [19, 20]. Since the seminal work of *McCammon*, *Gelin*, and *Karplus* demonstrating the dynamic nature of protein structures [19], numerous studies have appeared examining the behavior of proteins in the crystal lattice, *in vacuo*, and in aqueous solution [19]. These studies have shown that proteins possess a wealth of conformational substates of nearly equal energy [21]. The surrounding medium may influence protein flexibility by altering the energetic barriers separating these conformational substates and thereby affect the ease with which the protein may sample these states. *Hartsough* and *Merz* [18] provided one of the first molecular level pictures of a protein in a non-aqueous environment by simulating the behavior of bovine pancreatic trypsin inhibitor in water and chloroform *via* molecular dynamics calculations (MD). By this means the authors were able to evaluate the influence of the solvent upon protein structure and dynamics. After defining protein flexibility/rigidity of a protein as a measure of the

Table 1. Flexibility of BPTI (see Eq. 1) [18]

Run	$\langle RMS_{\text{all}}\rangle$/Å	$\langle RMS_{\text{back}}\rangle$/Å
H_2O, 298 K	0.741	0.531
$CHCl_3$, 298 K	0.528	0.418
$CHCl_3$, 310 K	0.614	0.476
$CHCl_3$, 360 K	0.630	0.512

amount by which instantaneous structures deviate from a time-averaged structure (Eq. (1)), the authors found that protein flexibility and solution temperature were intrinsically related.

$$flexibility = \langle\langle\langle\langle (r_{\mathrm{i}} - \langle r_{\mathrm{ave}}\rangle)^2\rangle\rangle^{1/2}\rangle \quad (1)$$

Bovine pancreas trypsin inhibitor (BPTI) in water as the solvent was calculated to be more flexible than BPTI in chloroform solution, even when the temperature of the chloroform solution was 50°C higher than that of the aqueous solution. (Table 1). Examination of the differential flexibility of each residue revealed that the increased flexibility of BPTI in water was distributed over several residues of the protein, the increase being greater for the terminal residue. Switching from water to chloroform strongly influenced the conformations of the amino acid side chains. In water, the side chains of the surface residues were almost fully extended into the solvent, whereas in chloroform the side chains were found to be folded back onto the surface of the protein. Consistent with this findings were the differences in the amount of solvent accessible surface area: using a common probe of 1.4 Å to facilitate comparisons, the solvent accessible surface area of BPTI was 4300 Å^2 in water and 3700 Å^2 in chloroform. From this evidence, it is apparent that the organic medium is unable to stabilize polar/charged side chains; these are folded back into the enzyme, thus reducing surface area. This observation that side chains do not contact with the organic solvent anymore was used to rationalize the increased thermostability of the protein in non-aqueous medium. In order for a protein to unfold, various hydrogen bond and salt bridge contacts must be disrupted. The ability to form hydrogen bonds with the solvent is important for protein unfolding. This will not be possible in the organic environment. This observation offered a good explanation to the noted increase of protein stability in nonpolar solvents [1, 3, 10, 22].

Furthermore, these results offer a possible explanation for "molecular imprinting" [12]. If the active site of the enzyme involves polar/charged residues, these side chains would be unable to reorient significantly in the organic environment. Once moved to their correct position by the inhibitor/imprinter, the polar residues would be unable to return to their old positions, as this would involve exposing these groups to the nonpolar solvent. This study was one of the first to provide atomic-level insights into the picture of a protein in a nonpolar environment. It showed that placing a protein in a non-aqueous environment results in a significant loss of conformational flexibility and a significant change in the side-chain conformations upon moving from aqueous to non-aqueous environment.

In another study on BPTI [23], *Hartsough* and *Merz* presented a detailed analysis using a MD approach of the dynamics of both protein and solvent, and the results

Table 2. Radius of gyration and *SASA* for BPTI [23]

Simulation conditions	$\langle r_{gyr} \rangle$/Å	*SASA* $\times 10^3$/Å^2
$CHCl_3$, 300 K	10.88	3.64
$CHCl_3$, 310 K	10.90	3.65
H_2O, 310 K	11.50	4.18

were compared to corresponding simulations in water. The results clearly demonstrated the profound effect the solvent environment can exert on protein structure and dynamics. In a polar aqueous environment, protein-solvent interactions obviously can successfully compete with protein-protein interactions. This is primarily true because of the charge separation present in the water molecule. The large partial charges on the hydrogen atoms of water present the opportunity for large-magnitude electrostatic interactions with the protein. By contrast, there is much less charge separation in the chloroform molecule, and the advantageous protein-solvent interactions are of a smaller magnitude. Accordingly, in a nonaqueous environment the protein must look for other sources of electrostatic stabilization. In chloroform the primary source of electrostatic stabilization available to charged side chains are the chloride counterions and other functional groups on the surface of the protein. The reduced mobility of the chloride ions was indicative of how small the intensity of solvent-charge interaction is in the chloroform-protein system. In water, both chloride ions and protein can establish more favorable interactions with each other and with the solvent. The conversion of protein-solvent interactions into protein-protein interactions has a substantial effect on enzyme properties. When protein-solvent contacts are replaced by protein-protein contacts, the enzyme decreases in size and increases in stability (Table 2). It has been pointed out that the greater thermal stability can be attributed to the cross-linking effect of protein-protein interaction. Although not as intense as chemically bonded cross-links, the strength of the electrostatic interactions will be substantial. Moreover, some distortions in the protein structure were observed, as indicated by the different amount of solvent accessible surface area.

Hydrogen bond network

An increasing network of protein-protein interaction should reduce the ability of the protein to unfold or undergo hinging motions. Simulations provided evidence that such motions will be inhibited by non-aqueous solvents. These results are also supportive of the experimental findings of higher rigidity/thermostability in organic solvents. The consequences of this intricated cross-linking network of hydrogen bonds was manifested in several experimental observations.

Merz proposed to use the above principles of cross-linking and intramolecular networking to design enzymes of even greater stability in organic solvents: the creation of salt bridges will preferentially stabilize the folded state. Specifically, if a pair of nonpolar residues, which are distant in the primary sequence, yet spatially proximate in the native state, are mutated to a positive/negative ion pair, a substantial increase in the stability of the folded state will result. In the fully solvent-exposed unfolded state, each of the charged residues will be relatively unstable in the nonpolar medium. Conversely, in the folded state an energetically favorable salt bridge will be formed by the ion pair. This new interaction will provide not only additional thermodynamic stability for the folded state but also an increased kinetic barrier to unfolding. This was also one of the first examples of an extensive use of MD simulations not only to study the behavior of enzymes in non-natural environments, but to provide insights and suggestions on how to improve protein molecules as catalysts.

In a recent paper, *Toba et al.* [24] have investigated the serine protease γ-chymotrypsin (γ-CT) in three different solvation environments using molecular dynamics. Unlike BPTI, γ-CT has been extensively used as a catalyst in several organic solvents for peptide synthesis and transesterifications [7], and the catalytic activities of serine protease have been well known for long. Moreover, the structure of the inhibited enzyme soaked in hexane as derived by X-ray diffraction was known [25, 26]. The simulations were: (1) γ-CT taken from the crystal structure of *Yennawar et al.* [25, 26] with seven surface bound hexane molecules and fifty essential water molecules all immersed in 1109 hexane molecules (simulation labeled CT); (2) γ-CT taken from Refs. [25, 26], solvated with fifty essential water molecules and immersed in 1107 hexane molecules (simulation CTWAT); and (3) γ-CT taken from Refs. [25, 26] solvated with a monolayer of 444 water molecules and immersed in 931 hexane molecules (simulation labeled CTMONO). From these trajectories the authors were able to find that the location of bound water molecules and the amount of hydration of the protein in the simulated structure had an effect on the protein flexibility as indicated by the changes in the root mean squared (*RMS*) deviation (Table 3). The radius of gyration value was similar for the three systems, indicating no significant unfolding or denaturation. Hydrophobic residues were found to have increased solvent accessible surface areas (*SASA*), whereas hydrophilic residues experienced a decrease in *SASA* in the CT and CTWAT simulations. No hexane diffusion into the interior of the protein was observed. Secondary structure analysis revealed that the active site structure was maintained in all simulations. Moreover, all fluctuations in positions and angles were within the scatter usually associated with protein fluctuations. This indicates that the active site of chymotrypsin remains relatively unchanged, which is in accordance with structural measurements of active enzymes in most organic solvents. The fluctuations in the simulation with water as the solvent were greater than those observed in the organic solvents. This observation corroborates the idea that proteins in non-aqueous solvents are less flexible; if the flexibility of the active site residues is critical for catalytic activity, this might partially explain the reduced activity of enzymes in organic solvents.

The simulations also confirmed previous findings that intramolecular forces such as hydrogen bonds and salt bridges were stronger in the CT and CTWAT simulations as shown by the increase in the number of stable hydrogen bonds. Net ion pair interactions and a reduced ratio of surface area to protein volume also contributed to the stability of the protein in hexane. The authors concluded that the stability of the protein in organic media is due to an increase in intramolecular stabilizing interactions such as hydrogen bonds, net ion pair interactions, and the reduced ratio of surface area to volume. The relatively hydrophobic protein interior

Table 3. Summary of the *RMS* deviation, radius of gyration, *SASA*, hydrophobic exposed *SASA*, and number of hydrogen bonds from the simulations on γ-chymotrypsin in three different solvation environments [24]

Analysis type	CT	CTWAT	CTMONO
	RMS deviation / Å		
Total	2.7	2.4	2.5
Backbone	2.4	2.1	2.0
	Radius of gyration / Å		
Minimized Structure	18.0	17.5	17.7
MD average	17.8	17.1	17.3
	$SASA \times 10^3$ / Å^2		
Crystal Structure	9.2	9.2	9.2
MD average	8.7	8.7	9.2
	Hydrophobic residues *SASA*/Total *SASA* (% exposed residues)		
Crystal Structure	34.8	34.3	34.3
MD Average	39.8	39.1	38.3
	Number of hydrogen bonds		
Crystal (total)	183	183	183
Min. Structure (total)	216	193	203
< 10%	216	224	294
> 90%	139	141	93

and the hydrophilic active site appear to be unaffected by the presence of the non-aqueous solvent.

In 1996, *Zheng* and *Ornstein* reported an MD simulation study of subtilisin Carlsberg in *DMSO* [27]. They compared their results to previous simulations carried out for water, acetonitrile, and carbon tetrachloride [28, 29]. *DMSO* is a dipolar, aprotic solvent with a dielectric constant of 46.5. It can compete with protein hydrogen bond acceptors for hydrogen bond donors. This is similar to acetonitrile, but *DMSO* is a much better hydrogen bond acceptor and, therefore, one of the few protein-dissolving solvents; and it has been used widely as an alternative solvent to water, due to the high solubility of peptides and proteins in it. *Zengh* and *Ornstein* also used quantum mechanics calculations to examine the hydrogen bonding interactions between *DMSO* and water and between *DMSO* and an amide hydrogen, using formamide as a model for the amide.

After analysis of their 745 ps trajectory they found that the structure of the protein was still not very different from the crystal structure in aqueous medium. However, some major differences could be noticed for DMSO, compared to other solvents. During the simulation for *DMSO*, the five residues of the N-terminus moved away and lost contact with the rest of the protein. This was in accordance with previous findings that *DMSO* is a protein denaturing medium. NMR studies by *Desai* and *Klibanov* also suggested that *DMSO* can cause partial unfolding of BPTI [30]. In addition, the simulation showed that *DMSO* was capable of stripping water molecules away from the protein surface, a phenomenon related to the hydrophilicity of the solvent. More water molecules were seen to leave the protein in *DMSO* than in acetonitrile, whereas in a hydrophobic solvent such as CCl_4 no water left the protein surface during the simulation. This was also in agreement with experimental observations [30]. It is worth noting that *DMSO* was found to be able to sequester a metal ion from the protein and coordinate it. The ability of polar solvents to sequester metal ions seems to be supported by the observation that when subtilisin crystals grown in water were soaked in acetonitrile, two metal ions were lost. This important observation provided an explanation to previous reports on the role of salts on protein stability and activity in organic solvents. The presence of salt ions can prevent direct contacts between organic solvents and the protein either by retaining more water molecules during lyophilization or by coordinating to solvent molecules to form complexes.

In these simulations [27–29] *Zheng* and *Ornstein* also found that the calculated number of intra-protein hydrogen bonds are different in *DMSO*, acetonitrile, and CCl_4 compared to water. The calculated average numbers of intra-protein hydrogen bonds were 247 ± 6, 259 ± 7, 264 ± 7, and 214 ± 7. Thus, the general trend that there are more intra-protein hydrogen bonds for a protein in a nonaqueous environment than in water seems to be followed once again. Through the application of quantum mechanics, the hydrogen bonding between *DMSO* and water was found to be stronger than the hydrogen bond between two water molecules, which was in agreement with experiment [27]. The calculations also showed that hydrogen bonding between *DMSO* and an amide hydrogen is stronger than between a carbonyl oxygen and an amide hydrogen [27], thus providing a good explanation why *DMSO* is a protein-dissolving solvent and is capable of inducing unfolding or deactivation of the enzyme.

Ornstein [31] also examined the effect of solvent environments on protein salt bridges by performing high level *ab initio* molecular orbital calculations in the gas phase and in three different solvents on a model compound represented by formate and guanidinium ions (Fig. 1). The energy difference between the neutral hydrogen-bonded complex and the zwitterionic form as well as the interconversion barrier between them was investigated in detail at 6-31G*, 6-311+G**, MP2/6-31G*, and MP2/6-311+G** levels. For calculations in solutions, the *Onsager* solvent reaction field model was used [32]. Three solvents (H_2O, *DMSO*, CCl_4) with different dielectric constants (78.3, 46.45, 2.238) were considered. In the gas phase, the neutral conventional hydrogen-bonded complex is predicted to be favoured at all four levels of theory. In a nonpolar, hydrophobic solvent like CCl_4, the energy difference between these two forms is small and the barrier that

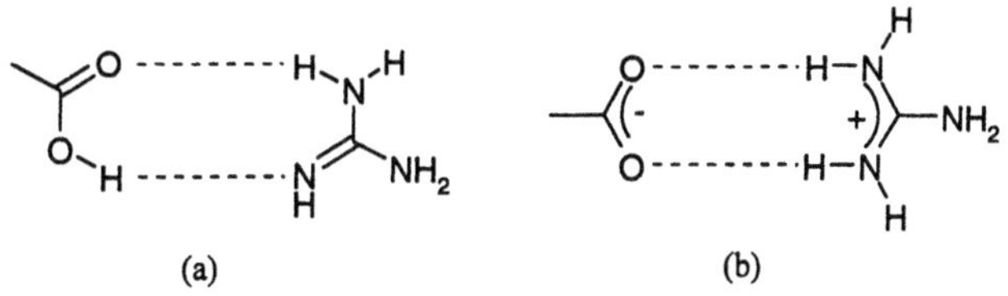

Fig. 1. Formate and guanidinium ion salt bridge: (a) neutral complex, (b) zwitterionic form [31]

Table 4. Calculated energy difference between the two forms of hydrogen-bonded complexes of formate and guanidinium (Figs. 1a and 1b) and the corresponding interconversion barriers [31]

Theory	Phase	Energy difference / $kJ \cdot mol^{-1}$	Forward barrier / $kJ \cdot mol^{-1}$
6-31 G	gas	22.6	33.1
6-311+G	gas	23.0	31.4
MP2/6-31 G	gas	14.2	16.3
MP2/6-311+G	gas	20.1	14.6
6-31 G	CCl_4	−0.4	22.6
6-311+G	CCl_4	0.8	20.9
MP2/6-31 G	CCl_4	−0.8	6.3
MP2/6-311+G	CCl_4	4.2	5.4
6-31 G	*DMSO*	−36.8	10.9
6-311+G	*DMSO*	−37.7	8.4
MP2/6-31 G	*DMSO*	−21.3	0
MP2/6-311+G	*DMSO*	−25.5	0
6-31 G	H_2O	−38.1	10.5
6-311+G	H_2O	−38.9	8.4
MP2/6-31 G	H_2O	−22.2	0
MP2/6-311+G	H_2O	−26.4	0

separates them is also low, but the neutral conventional hydrogen-bonded complex still seems to be slightly favoured. However, in polar solvents like *DMSO* and water, the zwitterionic form dominates (Table 4).

Ionized hydrogen bonds are often designated as salt bridges (or ion pairs) in crystallographic structures of proteins. In most cases, it is unknown on which atom the proton resides since X-ray crystallography cannot locate hydrogen atoms directly. In environments of low dielectric constant such as in non-aqueous solvents, membranes, and the interior of proteins, it is possible that the zwitterionic form will convert to the neutral hydrogen-bonded form *via* a proton shifting: calculations suggested that this really occurs in a very low dielectric environment [31]. Since a protein is not a homogeneous system with the same dielectric constant throughout, the microenvironment for each salt bridge could be different. Normally, salt bridges are also hydrogen-bonded to other polar groups or nearby water [33]; thus, the microenvironment for each salt bridge could be rather polar, and the effective dielectric constant could be larger than normally expected in the interior of a protein which is assumed to have a dielectric constant of 2–4 in most electrostatic models. Therefore, some of the interior salt bridges may still have significant zwitterionic character.

As seen before, MD simulations of proteins in non-aqueous solvents have shown that additional intra-protein hydrogen bonds and salt bridges are formed when a protein is transferred from an aqueous to a non-aqueous environment. Based on their calculations, *Zheng* and *Ornstein* [31] believe that a salt bridge is in the zwitterionic form in a relatively polar microenvironment. In a non-polar environment with very small dielectric constant it will probably prevail in the neutral form. Ref. [31] offers once more the suggestion that for a protein to have optimal activity in a nonpolar solvent isolated charged groups must be removed (by site-directed mutagenesis) in order to increase its stability. However, if a charged group has an oppositely charged group nearby to which it could form a salt bridge either in aqueous or in non aqueous solution, the interconversion between the neutral and zwitterionic forms *via* proton transfer is an effective way to annihilate charge separation.

Summarizing, by means of MD simulations several explanations can be given as to why enzymes have different structural characteristics in organic solvents compared to water: from all simulations a higher degree of intramolecular hydrogen bonding, improved salt bridge type interactions, and reduction of the surface area to volume ratio can be noticed in proteins and enzymes when soaked in organic solvents. All these factors contribute to the stability of the enzyme by preventing its possible unfolding through improving its intramolecular interactions. In addition, once unraveled, the amino acid side chains will experience less favorable contacts with the organic solvent which cannot hydrogen bond or electrostatically stabilize charged and polar side chains. Thus, the formation of the unfolded state can be considered to be more disfavoured in organic solvents than in water.

The higher number of intramolecular contacts can also be considered an important factor in determining the lower activity of enzymes in non-aqueous media: if the enzyme is, in fact, more rigid due to intraprotein interactions, it will be less able to adapt to the substrates entering the active site pocket. This phenomenon may be important as an explanation of the peculiar reactivity in organic solvents.

Modelling Enzyme Activity and Selectivity in Organic Media

A complete understanding of the enzyme-substrate-solvent interaction is necessary to increase enzyme utility to synthetic organic chemists [34, 35]. Several theories have been proposed to rationalize the mechanism by which organic media can influence reactivity.

In some cases, diffusional limitations have been claimed as the main cause of different and limited reactivity of enzymes in organic solvents. However, this possibility was ruled out by *Klibanov* and coworkers who have shown experimentally that slower catalysis is not due to mass transfer limitations [34].

Another hypothesis pointed out the change in conformation of the enzyme as the main cause of different reactivity compared to water. X-Ray crystallography experiments on crosslinked crystals of subtilisin soaked in both acetonitrile and dioxane showed that the three-dimensional structures of these proteins were virtually identical to their counterparts in water [36, 37]. Moreover, *Yennawar et al.* showed that uncrosslinked chymotrypsin in hexane had the same structure as in water [25]. By means of FT-IR spectroscopy, *Klibanov* and coworkers demonstrated that placing lyophilized subtilisin in organic solvents such as octane, acetonitrile, and dioxane had no appreciable effect on its secondary structure as indicated by its α-helix content [38].

The attention and the efforts of many researchers were then attracted to the study of the energy of binding between enzyme and substrate or to the study of the reaction intermediate (transition state) in the presence of both enzyme and solvent.

Thermodynamic approach and UNIFAC

A model proposed by *Klibanov* suggested that the selectivity dependence arises due to the differences in the thermodynamics of substrate solvation. Because this model has its basis in thermodynamics, it could make some quantitative predictions. One of the first applications involved the study of the substrate specificity of subtilisin Carlsberg in the transesterification reaction of N-Ac-*L-Ser*-OEt and N-Ac-*L-Phe*-OEt with 1-propanol in twenty different anhydrous solvents (Fig. 2). The serine substrate was strongly preferred in some solvents, whereas phenylalanine was preferred in others. The thermodynamic model represented specificity as a function of the solvent-to-water partition coefficients of the substrates and the substrate specificity of the enzyme-catalyzed hydrolysis of esters in water (Eq. (2)) [39].

$$\log((k_{cat}/K_M)_{Ser}/(k_{cat}/K_M)_{Phe})_{solvent} = \log(P_{Phe}/P_{Ser}) + \log((k_{cat}/K_M)_{Ser}/(k_{cat}/K_M)_{Phe})_{water} \quad (2)$$

Equation (2) predicts that a double-logarithmic plot of substrate specificity in any solvent *vs.* the P ratio will yield a straight line with a slope of 1, whose intercept with the ordinate should equal the substrate specificity in water. This model, while mechanistic and predictive, is essentially independent of the enzyme because the contribution of subtilisin-substrate binding is accounted for by the substrate specificity in water. Thus, this model should be applicable to any enzyme/substrate pair as long as the substrates are inaccessible to the solvent in the transition state.

However, a fundamental limitation of this approach is that it can be used only with water-immiscible solvents, because direct measurements of partition coefficients between water and water-miscible solvents are not feasible. In a second paper in 1993 those problems were overcome by introducing Eq. (3) [40] where γ is the thermodynamic activity coefficient of the substrate indicated.

$$P_{Phe}/P_{Ser} = (\gamma_{Ser}/\gamma_{Phe})_{solvent}/(\gamma_{Phe}/\gamma_{Ser})_{water} \quad (3)$$

The γ values for a given molecule in a solvent can be calculated on the basis of the *van der Waals* volumes and surface areas of the constituent groups of that molecule and of those of the solvent and from empirically determined interaction parameters

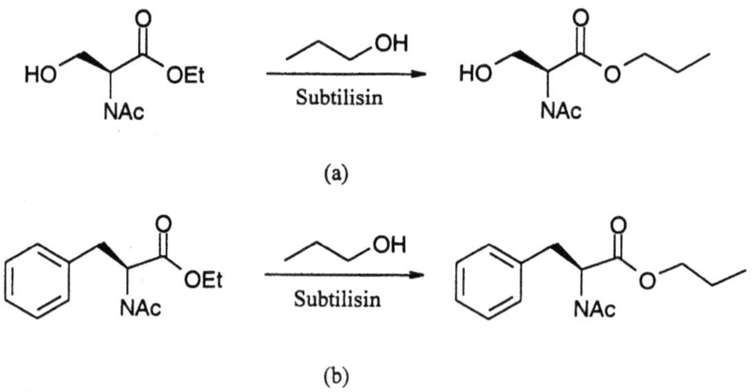

Fig. 2. Subtilisin-catalyzed transesterifications of N-Ac-*L-Ser*-OEt (a) and N-Ac-*L-Phe*-OEt (b) [39]

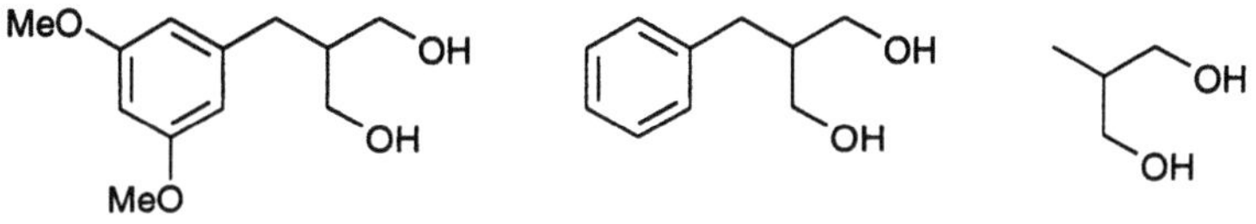

Fig. 3. 2-Substituted 1,3-propanediols [42]

between these groups. Such calculations can be carried out using the UNIFAC algorithm [41]. According to Eqs. (2) and (3) a double-logarithmic plot of substrate specificity *vs.* P ratios in several solvents should give a straight line as stated before. The authors were able to find a correlation coefficient of 0.96 for a line with a slope of 0.89 and an intercept of -1.7 (this was same as the experimental value of the log of the substrate specificity in water). Once again, the assumption that substrates be fully removed from the solvent in the transition state is made, which cannot be always true. Thus, *Klibanov* and coworkers devised a new thermodynamic method to rationalize the solvent dependence of any type of enzymatic selectivity (prochiral, diastereo-, or enantioselectivity) solely on the basis of the thermodynamics of substrate solvation [42]. The model predicts that selectivity (defined as the ratio k_{cat}/K_M) should be proportional to the ratio of the thermodynamic activity coefficients of the desolvated portions of the substrates in the relevant transition state of the enzymatic reaction. This ratio is calculated by a three step procedure: (*a*) determine the desolvated portion of the substrate in the transition state using molecular modeling based on the crystal structure of the enzyme, (*b*) approximate the desolvated portion of the substrate by a distinct model compound, and (*c*) calculate the activity coefficients of this model compound using the UNIFAC algorithm.

In the first study [42] the model was applied to the determination of the prochiral selectivity of enzymes, in particular γ-chymotrypsin and subtilisin Carlsberg used as asymmetric catalysts in the acetylation of 2-substituted 1,3-propanediols (Fig. 3). The relation on which the method is based is represented by Eq. (4).

$$\log((k_{cat}/K_M)_I/(k_{cat}/K_M)_{II})A = \log(\gamma_I/\gamma_{II}) + \log((k_{cat}/K_M)_I/(k_{cat}/K_M)_{II})B \quad (4)$$

In Eq. (4), I and II represent either the two transition state models for prochiral selectivity or for enantioselectivity. Moreover, if solvent B is fixed as a reference solvent, the final term in Eq. (4) can be taken as a constant, and the equation will assume the simplified form of Eq. (5).

$$\log((k_{cat}/K_M)_I/(k_{cat}/K_M)_{II})_A = \log(\gamma_I/\gamma_{II}) + const \quad (5)$$

From a theoretical point of view, the authors demonstrated that the selectivity depends only on the differences in the desolvation of the transition state. This is the point where molecular modelling comes into play: MD is, in fact, used to determine the desolvated portions of the two substrates leading to two different products. In the case of prochiral selectivity, very good results were obtained in the case of cross-linked crystal enzyme, where a correlation factor of 0.94 and a slope of 1 for the double-logarithmic plot of Eq. (5) was obtained.

Very good regression coefficients (~0.93) were also obtained in the case of the enantioselectivity in the transesterification reaction of methyl 3-hydroxy-2-phenylpropionate in different organic solvents [43]. In this paper, however, the authors pointed out how the approximation of the desolvated part of the substrate can be made using different "UNIFAC groups". Moreover, it was shown that the predictivity degree of the method was heavily dependent on the choice of these approximating groups.

A Monte Carlo/Energy Minimization scheme was also used to determine the desolvated part of the transition intermediate model, and then the UNIFAC procedure was applied [44]. In this case, the UNIFAC approach proved to be able to give qualitative predictions, as well, but was unable to afford good quantitative correspondence between experimental and calculated values. *Luque et al.* applied this approach to the study of the role of transition state desolvation on enzymatic enantioselectivity in aqueous-organic mixtures using lipases as catalysts. The purely thermodynamic treatment resulted in very poor predictions of enantioselectivity [45]. A thorough analysis revealed that the energetics of substrate desolvation did not contribute significantly to the experimentally observed changes in the enantioselectivity of lipases. Therefore, other factors such as conformational changes in enzyme molecules induced by the solvents or differential solvent displacement from the active center of the enzyme by the substrate were claimed to be fundamental [45]. Thus, a more in-depth computational study was required to evaluate the influence of the solvent on enzyme reactivity, based on force-field or higher level studies of the interactions between the substrate, the enzyme, and, possibly, the solvent.

Transition state and molecular dynamics

Hult and coworkers [46] used information from the crystal structures of the *Candida antarctica* lipase type B (CALB; incubated with the detergent Tween 80, or inhibited by a racemic mixture of *n*-hexylchlorophosphonate ethyl ester) to model the tetrahedral intermediates of two acyl transfer reactions into the active site in order to explain the high selectivity of CALB towards certain secondary alcohols. The mechanism was assumed to be identical to the general one of serine-proteases or hydrolases. The first intermediate model represented the transesterification reaction of 1-phenylethanol with an octanoate ester as acyl donor. In the second intermediate, the hydrolysis of a glycerol butanoate derivative was studied (Figs. 4 and 5). Molecular dynamics and minimization techniques were used throughout. Starting models of the enzyme-substrate transition states were built as tetrahedral covalent complexes with the Oγ atom of the active site serine residue connected to the carbonyl carbon atom of the substrate. A protonated histidine residue in the catalytic triad was assumed to be present in the transition state. In this reaction mechanism, the protonated active site histidine forms hydrogen bonds to the Oγ atom of the catalytic serine and to the oxygen of the alcohol group of the substrate. Furthermore, the oxyanion of the transition state was forced to interact with the oxyanion hole residues. The energies and structures of the complexes were calculated by a combined energy minimization and molecular dynamics protocol [46]. Energy minima of the complexes were found by MD simulations *in vacuo* in

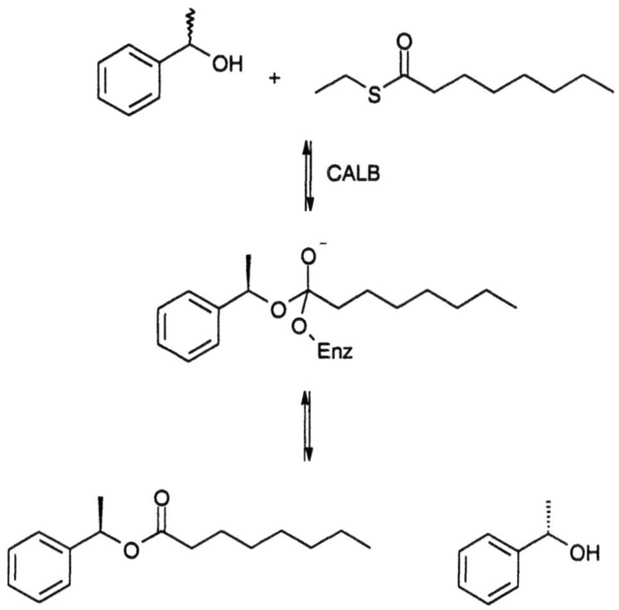

Fig. 4. Stereoselective resolution of 1-phenylethanol with CALB and thioethyl-octanoate [46]

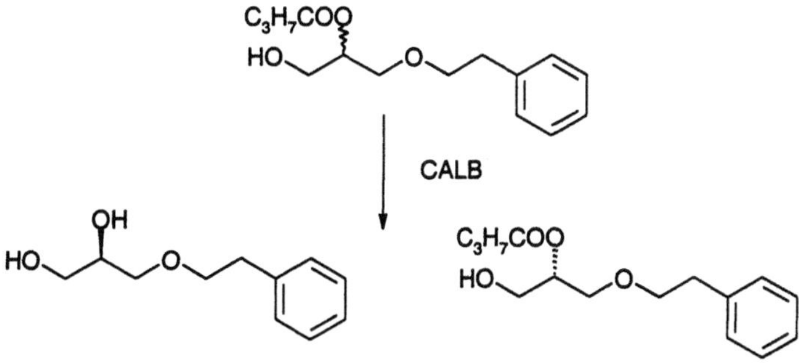

Fig. 5. CALB catalyzed stereoselective hydrolysis of glycerol butanoate derivative [46]

which the molecules were repeatedly heated to 300 K and slowly cooled to 1 K. The trapped conformations were then further minimized with 500 steps of energy minimization.

In the case of 1-phenylethanol the (*S*)-alcohol was found in such a conformation that it could not easily donate a proton to the catalytic hystidine 224 residue, which is considered to be essential for nucleophilic attack on the acyl enzyme. Consequently, in a hydrolysis reaction, the alcohol moiety of the substrate would be stuck in a position where it cannot accept a proton from the histidine and is therefore not efficient as a leaving group. On the other hand, the energy minima of the tetrahedral intermediate of the (*R*)-enantiomer contain a favourable hydrogen bond between the catalytic histidine and the oxygen atom of the alcohol group which would enable rapid proton transfer for catalysis.

In the glycerol derivative hydrolysis study, the fast reacting (*R*)-enantiomer was found to have all its bonds in the carbon chain of the large group in the relaxed *trans* conformations. The small group was interacting with *Trp*104 in the small pocket. On the other hand, the slow reacting (*S*)-enantiomer had its large group twisted away from the specificity pocket, in contact with the side chains of *Leu*278, *Ala*282, and *Ile*285 side chains. The presence of these side chains forced the carbon chain of this group of the substrate to adopt an energetically less favorable conformation. It was hypothesized that the enantiopreference for the (*R*)-enantiomer in the hydrolysis of glycerol butanoate was due to these strain differences in the substrate conformations. The results of this study (together with X-ray data) provided structural explanations for the high stereoselectivity of CALB toward secondary alcohols which can be extended to more general cases. However, all MD and energy minimization calculations presented in Ref. [46] were run *in vacuo* and gave only structural-sterical information about interactions between the enzyme and the tetrahedral intermediate.

Semiempirical and ab initio calculations

Haeffner et al. also proposed an MD-based molecular modelling method to predict the enantioselectivity in lipase-catalyzed transesterification reactions of the substrates of Fig. 6 [47]. A tetrahedral intermediate with a rigid central part was used to mimic the transition state. The geometric parameters were calculated by means of *ab initio* methods on the tetrahedral intermediate formed in the gas phase reaction between methyl acetate and a methoxide ion using the 6-31 + G^* basis set. The charges were calculated with an electrostatic potential (ESP) fitting methodology on the tetrahedral intermediate using 6-31 + G^* on RHF/STO-3G optimized geometry. The same set of charges was used for both enantiomers in all the three cases. After MD conformational steps and energy refinement of the obtained structures, enantioselectivity was expressed as a function of the energy differences between the two diastereoisomeric enzyme-substrate complexes. This energy difference evaluation was performed by defining subsets within the enzyme structures. Two different strategies were used. The first one used predefined parts of the enzyme and the substrate as subsets, whereas the second approach formed energy based subsets. The selection of the residues to be included was based on the energy of interaction between the specific residue and the transition state analogue. Importantly, this approach was able to predict which enantiomer was the fast-reacting one.

However, the issue of solvent influence, along with that of the electrostatic influence of the enzyme on reactivity, was not examined. This problem could be tackled by the use of mixed quantum mechanics/molecular mechanics (QM/MM) methods which were first introduced by *Warshel* and *Lewitt* [48]. Using this

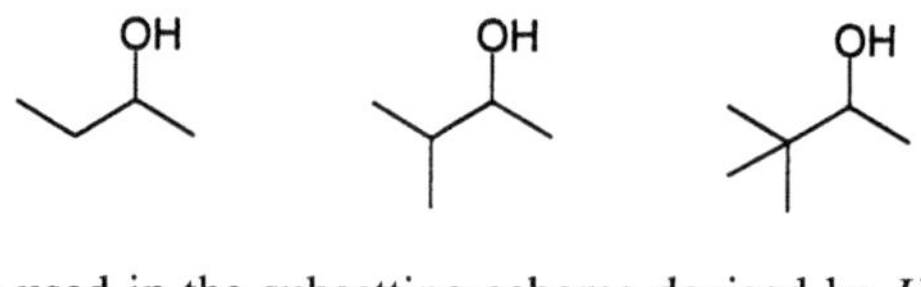

Fig. 6. Substrates used in the subsetting scheme devised by *Haeffner et al.* [47]

approach, the reactive events can be described at a QM level for the active site, whereas the rest of the protein is treated by MM or MD calculations. The two regions can interact through an interaction Hamiltonian; in this way, the active site (*i.e.* the QM region) can respond to the influence of its protein and solvent environment in terms of polarization, electron correlation, and charge transfer effects.

In 1991 *Kollman* and coworkers described semiempirical molecular orbital calculations of the serine protease catalyzed hydrolysis of amides and esters and the effect of the protein environment and dynamics on the process [49]. They found that the PM3 Hamiltonian was much better suited than the AM1 one at reproducing hydrogen bond geometries, and so the former was used to describe the reaction mechanism. The lowest energy pathway for formation of the tetrahedral intermediate was for serine to approach the substrate, followed by coupled heavy atom motion and proton transfer to complete the reaction. The importance of active site residues and the environment in stabilizing the substrate was addressed by evaluating particular interactions in fully solvated enzyme-substrate models. These interaction energies were calculated for both a noncovalent enzyme-substrate complex and a model for the tetrahedral intermediate in which a covalent bond was formed between the serine and substrate. The oxyanion hole and *Asp* greatly stabilized the active site region of the tetrahedral intermediate and, presumably, the transition state structure. The environment itself, excluding the active site residues and the oxyanion hole, was important in stabilizing the scissile bond of the substrate in the tetrahedral intermediate *vs.* the *Michaelis* complex. This was certainly one of the first important reports of the use of information gained from quantum mechanical simulations in elucidating mechanistic pathways in which the role of the protein environment was not negligible.

QM/MM approaches were used successfully to describe the structural and reactivity characteristics of metal containing enzymes, to design a charge model to describe the active site of these enzymes, and, finally, to describe the binding preferences of inhibitors to Zn(II) containing proteins [50].

Williams and coworkers were able to describe and characterize the transition structure for the reduction of pyruvate, catalyzed by lactate dehydrogenase (LDH), by means of QM/MM calculations involving a fully flexible active region comprising 1900 atoms and 5700 degrees of freedom [51].

Another successful application of this technique is due to *Antonczack et al.* who described the thermolysin catalyzed hydrolysis of formamide [52]. The first step of the reaction was proven to be the nucleophilic attack of the carbon atom by the oxygen of the water molecule or by a water dimer. The mechanism involving an ancillary water molecule is always favoured compared to the process in which a single water molecule is involved. The fact was explained by a better nucleophilicity of the oxygen atom in the water dimer. The zinc atom of the catalytic center acts as a *Lewis* acid and the ligands as electron reservoirs. To reach these conclusions, the authors needed an accurate description of the active site of the enzyme, taking into account the metal and the environment. In particular, the ligands of Zn(II) were seen to play an important role in delocalizing the charges along the reaction pathway. Thus, this study was feasible thanks to the possibility of describing the active site region at a good semiempirical level (AM1) and the rest of the molecule through a classical force field.

A striking application of this methodology to describe enzyme reactivity is due to *Warshel*'s group [53]. They developped an effective approach for *ab initio* calculations of activation energies in enzymatic reactions. Their approach used an empirical valence bond (EVB) potential surface as a reference potential for evaluating the free energy of a hybrid *ab initio* QM/MM potential surface. This procedure involved an automated calibration of the EVB potential using gas-phase *ab initio* calculations. The method was used to study the nucleophilic attack step in the catalytic reaction of subtilisin. It was found that the use of the EVB potential as a reference allowed the determination of the actual activation free energy of enzymatic reactions [53].

Amide hydrolysis in trypsin and water solution was studied by *Stanton et al.* [54]. They proposed a new more general method to combine *ab initio* quantum mechanical calculations with the classical mechanical free energy perturbation approach to calculate the energetics of enzyme-catalyzed reactions and the same reactions in solution. This method, enabling enzyme and solution reactions to be compared without the use of empirical parameters, was applied to the formation of the tetrahedral intermediate in trypsin. Two critical aspects of this new approach were: (*a*) the calculation of the reaction energetics in solution and (*b*) the use of the restrained electrostatic potential (RESP) protocol [55] to calculate the charge distributions of structures along the reaction pathway. This enabled the authors to circumvent the problem of partitioning the charge across a residue divided into QM and MM parts. These charges were then used along with standard parameters to define the force field representations of the complex in the free energy perturbation (FEP) calculations. Two sets of calculations were performed to obtain the differences in free energy between points on the reaction pathway. In the first set, the QM region was held static during the entire length of the simulation. In the second set of simulations, internal degrees of freedom of the QM region were allowed to relax in order to respond to the dynamics of the enzyme. In this way, the impact of active site flexibility on free energy calculations could be assessed. This protocol allows to examine the nature of enzyme catalysis by performing analogous simulations in water. Comparison of these two simulations yields insights into the role of the enzyme as a catalyst. The method is thus certainly general and applicable to the investigation of the energetics of enzyme reaction pathways.

Continuum and explicit solvents

The use of *ab initio* calculated charges to describe enantioselectivity was described by *Ke*, *Tidor*, and *Klibanov* [56]. The transition states for the acylation and deacylation of chymotrypsin by substrates **1–4** Fig. 7 were approximated by the corresponding transition intermediates. The substrate charge distribution for the (*R*) and (*S*) configured substrates was determined *via* STO-3G electrostatic potential fits using an ensemble of ten energy-minimized structures obtained from gas-phase MD simulations. The substrates were then docked into the enzyme active site, and vacuum (modelled using a distance-dependent dielectric) MD simulations followed by energy minimization were used to obtain suitable structures of the enzyme/substrate complexes. To mimic the presence of water, the authors used continuum electrostatic models, and the $\Delta\Delta G'$ activation energy

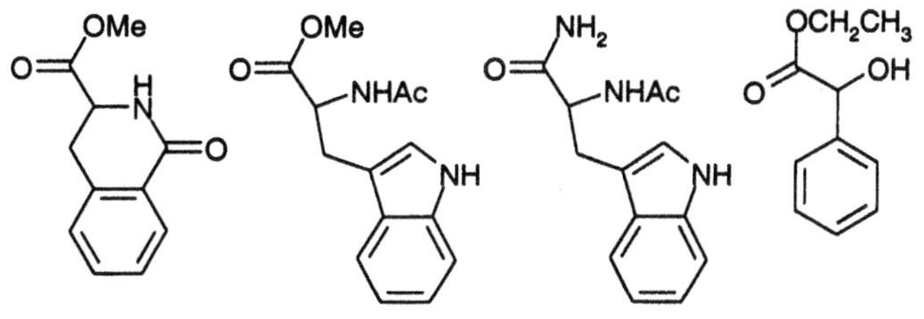

1 2 3 4

Fig. 7. Substrates used by *Ke et al.* as reported in Table 5 [56]

Table 5. Comparison of experimentally determined enantioselectivities in the chymotrypsin catalyzed hydrolyses of substrates **1–4** of Fig. 7 with the simulation results *in vacuo* and in aqueous solution [56]

	$\Delta\Delta G' /$ kj · mol^{-1}		
		Calculated	
Substrate	Experimental In buffer	*in vacuo*	In water
1	−20.5	−18.4	−18.8
2	>30	26.4	22.2
3	>30	33.5	36.4
4	0.4	5.0	0.4

differences in water between (*R*) and (*S*) substrates were calculated according to Eq. (6) where the first term is the difference in the bonded energy terms and the second is the difference in the *van der Waals* energy for molecular mechanics energy-minimized structures of the two intermediates. The two ΔG values in the second term of the equation are the electrostatic contributions in continuum solvent consisting of coulombic and reaction field contributions, whereas the nonpolar term is computed from the solvent-accessible surface area difference between the (*R*) and (*S*) configured enzyme-bound tetrahedral intermediates. The results were in good agreement with experimental values as reported in Table 5.

$$\Delta G_{R-S}^{\text{water}} = \Delta E_{R-S}^{\text{covalent}} + \Delta E_{R-S}^{vdW} + \Delta G_{R-S}^{\text{elec}} + \Delta G_{R-S}^{\text{nonpolar}} \quad (6)$$

This study was a considerable advance over earlier efforts, but significant improvements could still be made, especially through the use of explicit solvent and the determination of atomic point charges including the effect of the enzyme environment. This had been done in the past on ester cleavage by a β-cyclodextrin, but the first application to enzyme systems is due to *Colombo et al.* [57]. Their study focussed on the development of a deeper understanding of the enantioselectivity of the serine protease subtilisin *DMF* in with respect to the resolution of a racemic mixture of 1-phenyl-ethanol by a transesterification reaction with the acylating agent vinyl acetate. Once again, the formation of the tetrahedral intermediate is thought to be the rate determining step for catalysis (Fig. 8). Moreover, it is supposed that the structure of the transition state for the formation

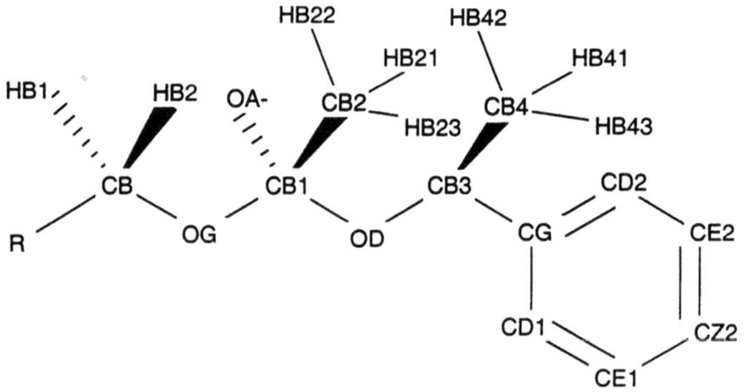

Fig. 8. Tetrahedral intermediate for the subtilisin-catalyzed transesterification of 1-phenyl ethanol with acetate (see Table 6) [57]

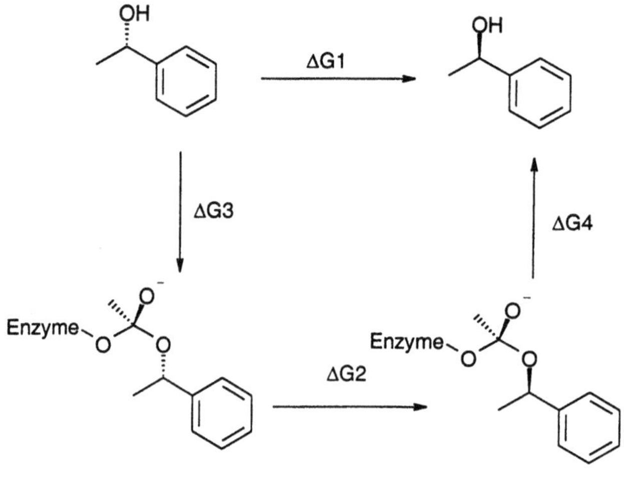

Scheme 1. Scheme for FEP calculation

of the tetrahedral intermediate closely resembles the structure of the tetrahedral intermediate itself. The behavior and the characteristics of the tetrahedral intermediates for both the (*R*) and (*S*) enantiomers were examined by means of MD and FEP simulations (Scheme 1) with an explicit representation of the organic solvent *DMF*. A critical aspect of this study was the determination of the charge distribution of the (*R*) and (*S*) tetrahedral intermediates by means of a QM/MM electrostatic potential fitting methodology. In designing the active site model, the (*R*) and (*S*) configured tetrahedral intermediates were found to have significantly different charge distributions due to the presence of the stereo-differentiating environment presented by the enzyme. In contrast, the charge distribution obtained for models of the tetrahedral intermediate in the gas-phase display similar charge distributions. From MD simulations, both steric and, more importantly, electrostatic complementarity were seen to play a role in the determination of the enantioselectivity of this enzyme reaction. Table 6 displays the differential charges on the tetrahedral intermediates for the two enantiomers and on the active site residue atoms. These point charges were then used for FEP simulations to obtain a free energy difference that was in good accordance with experiment (Table 7),

Table 6. Average ESP charges calculated for the (*R*) and (*S*) complexes of Fig. 8 [57]

	(*R*)-complex		(*S*)-complex		
Atom	Charge	Std. dev.	Charge	Std. dev.	Charge diff.
CB	−0.056	0.062	−0.029	0.063	−0.027
HB2	0.081	0.027	0.064	0.029	0.017
HB3	0.035	0.023	0.042	0.036	−0.007
OG	−0.305	0.061	−0.351	0.033	0.047
CB1	0.588	0.106	0.697	0.101	−0.108
CB2	−0.331	0.065	−0.297	0.067	−0.034
HB21	0.075	0.038	0.045	0.033	0.030
HB22	0.092	0.027	0.046	0.026	0.046
HB23	0.060	0.027	0.063	0.034	−0.003
OA-	−0.876	0.036	−0.914	0.038	0.037
OD	−0.462	0.073	−0.543	0.065	0.081
CB3	0.205	0.118	0.464	0.118	−0.258
HB3	0.045	0.059	−0.044	0.038	0.089
CB4	−0.277	0.051	−0.222	0.059	−0.054
HB41	0.052	0.022	0.042	0.027	0.010
HB42	0.065	0.024	0.033	0.031	0.032
HB43	0.095	0.027	0.050	0.017	0.045
CG	0.055	0.071	−0.047	0.075	0.103
CD1	−0.154	0.048	−0.069	0.053	−0.085
HD1	0.095	0.020	0.126	0.019	−0.031
CE1	−0.109	0.054	−0.142	0.049	0.033
HE1	0.101	0.021	0.126	0.021	0.025
CZ	−0.114	0.052	−0.116	0.042	0.001
HZ	0.097	0.020	0.093	0.023	0.004
CE2	−0.107	0.037	−0.131	0.034	0.024
HE2	0.111	0.014	0.100	0.011	0.011
CD2	−0.125	0.051	−0.139	0.045	0.014
HD2	0.117	0.013	0.099	0.015	0.018

Table 7. Calculated free energy difference for the thermodynamic cycle given in Scheme 1 [57]

Starting structure	FEP time / ps	Free energy ($\Delta\Delta G'$)/kJ $\cdot$ mol^{-1}
90 ps	150	6.3 ± 2.1
90 ps	450	2.5 ± 1.2
90 ps	750	5.0 ± 2.1
120 ps	450	3.3 ± 1.7
120 ps	750	5.4 ± 2.1

quantitatively supporting the accuracy of the model and suggesting that all-atom molecular simulations are capable of providing accurate qualitative and quantitative insights into enzyme catalysis in non-aqueous solvents as well as into enzyme catalysis in general.

Conclusions

In this article we have reviewed how molecular modeling techniques have been used in the last few years to shed some light on how organic solvents influence the molecular characteristics of proteins and enzymes. MD simulations on BPTI, chymotrypsin, and subtilisin made it possible to get a deeper understanding into how increased intramolecular interactions improve conformational rigidity, thus explaining the lower reactivity and the higher thermostability of enzymes in non-aqueous media. This structural information allowed to understand some peculiar aspects of the behavior of enzymes in organic solvents.

The application of thermodynamics-based models allowed the first qualitative predictions on the selectivity of many reaction types. However, only after the application of quantum mechanical/molecular mechanical methods quantitative models of actual reactivity patterns could be realistically formulated. Even though these methods are still in part approximate due to the partial application of force fields and semiempirical *Hamiltonians*, they showed that focusing the attention on the reactive region of the enzyme with QM calculations, while keeping track of the enzyme-environment influence through MD, was the right way to go. Improvements in the QM computational methods, such as the introduction of linear scaling divide-and-conquer methods [58], will certainly allow to gain a first principle understanding of the reactivity of macromolecules in any kind of environment.

Acknowledgements

We thank the CNR Target Project on Biotechnology for financial support.

References

[1] Klibanov AM, Zacks A (1984) Science **224**: 1249
[2] Klibanov AM (1986) CHEMTECH **63**: 354
[3] Zacks A, Klibanov AM (1986) J Am Chem Soc **108**: 2767
[4] West JB, Hennen WJ, Lalonde JL, Bibbs JA, Zhong Z, Meyer EF jr, Wong CH (1990) J Am Chem Soc **112**: 5313
[5] Zacks A, Klibanov AM (1985) Proc Natl Acad Sci USA **82**: 3192
[6] Kuhl P, Halling PJ, Jakubke HD (1990) Tetrahedron Lett **31**: 5213
[7] Klibanov AM (1989) TIBS **14**: 141
[8] Carrea G, Ottolina G, Riva S (1995) Trends Biotechnol **13**: 63
[9] Kazandjian RZ, Klibanov AM (1986) J Am Chem Soc **107**: 5448
[10] Ayala G, De Gomez-Poyou MT, Gomez-Poyou A, Darszon A (1986) FEBS **203**: 41
[11] Capaldi RA, Malatesta F, Darley-Usmar VM (1983) Biochim Biophys Acta **726**: 135
[12] Russel A, Klibanov AM (1988) J Biol Chem **263**: 11624
[13] Stahl M, Jeppsson-Wittstrand U, Mannson MO, Mosbach K (1991) J Am Chem Soc **113**: 9366
[14] Poole PL, Finney JL (1983) Biopolymers **22**: 255
[15] Poole PL, Finney JL (1983) Int J Macromol **5**: 308
[16] Poole PL, Finney JL (1984) Comments Mol Cell Biophys **2**: 129
[17] Careri G, Gratton E, Yang PH, Rupley JA (1980) Nature **284**: 572
[18] Hartsough DS, Merz KM (1992) J Am Chem Soc **114**: 2570
[19] McCammon JA, Gelin BR, Karplus M (1977) Nature **267**: 585

[20] Brooks CL III, Karplus M, Pettitt BM (1988) Proteins: A Theoretical Perspective of Dynamics, Structure, and Thermodynamics. Wiley, New York
[21] Frauenfelder H, Sligar SG, Wolynes PG (1991) Science **254**: 1598
[22] Zacks A, Klibanov AM (1988) J Biol Chem **263**: 3194
[23] Hartsough DS, Merz KM Jr (1993) J Am Chem Soc **115**: 6529
[24] Toba S, Hartsough DS, Merz KM Jr (1996) J Am Chem Soc **118**: 6490
[25] Yennawar NH, Yennawar HP, Farber GK (1994) Biochemistry **33**: 7326
[26] Yennawar NH, Yennawar HP, Farber GK (1995) J Am Chem Soc **117**: 577
[27] Zheng Y-J, Ornstein RL (1996) J Am Chem Soc **118**: 4175
[28] Zheng Y-J, Ornstein RL (1996) Biopolymers **38**: 791
[29] Zheng Y-J, Ornstein RL (1996) Protein Eng **9**: 485
[30] Desai UR, Klibanov AM (1995) J Am Chem Soc **117**: 3940
[31] Zheng Y-J, Ornstein RL (1996) J Am Chem Soc **118**: 11237
[32] Onsager L (1936) J Am Chem Soc **58**: 1486
[33] Hwang J-K, Warshel A (1988) Nature **334**: 270
[34] Klibanov AM (1997) Trends Biotechnol **15**: 97
[35] Tramper JVM, Vermüe MH, Beeftink HH, Von Stockar U (1992) Biocatalysis in Non-Conventional Media. Progress in Biotechnology, vol 8. Elsevier, Amsterdam
[36] Fitzpatrick PA, Steinmetz ACU, Ringe D, Klibanov AM (1993) Proc Natl Acad Sci USA **90**: 8653
[37] Fitzpatrick PA, Ringe D, Klibanov AM (1994) Biochem Biophys Res Comm **198**: 675
[38] Griebenow K, Klibanow AM (1997) Biotechnol Bioeng **54**: 210
[39] Wescott CR, Klibanov AM (1993) J Am Chem Soc **115**: 1629
[40] Wescott CR, Klibanov AM (1993) J Am Chem Soc **115**: 10362
[41] Fredenslund A, Gmelhling J, Rasmussen P (1977) Vapor-Liquid Equilibria Using UNIFAC. Elsevier, Amsterdam
[42] Ke TW, Klibanov AM (1996) J Am Chem Soc **118**: 3366
[43] Wescott CR, Klibanov AM (1996) J Am Chem Soc **118**: 10365
[44] Colombo G, Ottolina G, Carrea G, Bernardi A, Scolastico C (1998) Tetrahedron Asymm **9**: 1205
[45] Luque S, Ke T, Klibanov AM (1998) Biocatal Biotransform **16**: 233
[46] Uppenberg J, Ohrner N, Norin M, Hult K, Kleywegt GJ, Patkar S, Waagen V, Anthonsen T, Jones TA (1995) Biochemistry **34**: 16838
[47] Haeffner F, Norin T, Hult, K (1998) Biophysical J **74**: 1251
[48] Warshel A, Levitt M (1976) J Mol Biol **103**: 227
[49] Daggett V, Schröder S, Kollman P (1991) J Am Chem Soc **113**: 8926
[50] Merz KM Jr, Banci L (1996) J Am Chem Soc **118**: 10235
[51] Turner AJ, Moliner V, Williams IH (1997) J Chem Soc Chem Commun 1271
[52] Antonczack S, Monard G, Ruiz-Lopez MF, Rivail J-L (1998) J Am Chem Soc **120**: 8825
[53] Bentzien J, Muller RP, Florian J, Warshel A (1998) J Phys Chem B **102**: 2293
[54] Stanton RV, Perakyla M, Bakowies D, Kollman PA (1998) J Am Chem Soc **120**: 3448
[55] Cornell WD, Cieplak P, Bayly CI, Gould IR, Merz KM Jr, Ferguson DM, Spellmeyer D Fox T, Caldwell JW, Kollman PA (1995) J Am Chem Soc **117**: 5179
[56] Ke T, Tidor B, Klibanov AM (1998) Biotechnol Bioeng **57**: 741
[57] Colombo G, Toba S, Merz KM Jr (1999) J Am Chem Soc **121**: 3486
[58] Dixon SL, Merz KM Jr (1996) J Chem Phys **104**: 6643

Received November 18, 1999. Accepted (revised) February 8, 2000

Invited Review

Lipase-Catalyzed Synthesis of Carboxylic Amides: Nitrogen Nucleophiles as Acyl Acceptor

Fred van Rantwijk*, **Michiel A. P. J. Hacking,** and **Roger A. Sheldon**

Laboratory of Organic Chemistry and Catalysis, Delft University of Technology, NL-2628 BL Delft, The Netherlands

Summary. The lipase-catalyzed aminolysis of carboxylic esters is a fairly general reaction that has been performed with a wide range of esters and amines, generally in anhydrous organic media to avoid undesirable hydrolysis of the ester. Alternatively, carboxylic amides can be synthesized by lipase mediated condensation of carboxylic acids and amines if an excess of either reactant is avoided.

Chiral carboxylic esters have been resolved by lipase-catalyzed aminolysis. In the majority of these resolutions, *Candida antarctica* lipase B has been employed as the catalyst. A range of chiral amines has been resolved by lipase mediated acylation, using mainly the lipases from *C. antarctica* (B type) and *Pseudomonas* species. The enantiorecognition was frequently found to depend critically on the acylating agent and the reaction medium.

Keywords. Lipase; *Candida antarctica* lipase; Amine; Carboxylic amide; Enantioselectivity.

Introduction

There is currently a marked trend in the fine chemicals industry towards the reduction, or preferably elimination, of waste by widespread substitution of classical syntheses employing stoichiometric reagents with cleaner catalytic alternatives [1].

Enzymes occupy a special position in the catalysis field because they operate at moderate temperature and mild reaction conditions, which is advantageous with sensitive reactants and products. Moreover, enzymes are inherently enantioselective. In consequence, enzyme catalysis is destined to play a major role in the transition from stoichiometric to catalytic transformations, particularly if it can be made to transcend the limitations of natural reactions in aqueous medium by replacement of the natural reactant by a non-natural one. Lipases, the lipid hydrolyzing catalysts of nature, are eminently suited for such an approach because of their stability and the simplicity of their catalytic machinery (Fig. 1). These characteristics have made them outstanding catalysts for synthetic biotransformations that involve the carboxyl group, such as esterification, transesterification,

* Corresponding author

perhydrolysis, and aminolysis in which the natural nucleophile (water) is replaced by alcohols, hydroperoxides, or amines [2, 3]. The principle of lipase-catalyzed aminolysis, resulting in the synthesis of carboxylic amides, has been demonstrated for the first time in 1984/85 [4, 5], and a multitude of publications describing its application has been published since.

This review describes the use of nitrogen nucleophiles in the acyl accepting role as a synthetic route to carboxylic amides as well as the resolution of chiral carboxylic acids, esters, and amides by lipase catalyzed aminolysis.

Mechanistic Considerations

Lipases, the lipid splitting catalysts of nature (EC 3.1.1.3), operate by essentially the same mechanism as that of the serine proteases [6]. The active site (see Fig. 1) contains a serine residue that is activated by histidine and aspartate residues; these together form the catalytic triad. The reactant ester forms a tetrahedral acyl-enzyme intermediate by reaction with the OH group of the catalytic serine residue; the resulting excess of negative charge that develops on the carbonyl oxygen atom is stabilized by the oxyanion hole. Next, the tetrahedral intermediate collapses to the serinate ester with elimination of the alcohol. Subsequent reaction of the acylenzyme intermediate with the nucleophile – the acyl acceptor – affords the product. In the case of hydrolysis, which is the natural reaction of lipases, the nucleophile is water. However, almost any nucleophile can, in principle, react with the acyl-enzyme intermediate [2].

Ester aminolysis can be regarded as irreversible, because the transformation is strongly exothermic under normal reaction conditions [7]. Amide hydrolysis, which is, *inter alia*, not readily mediated by lipases, can also be excluded due to the almost universally employed anhydrous reaction conditions (see below). Lipase mediated aminolysis of a carboxylic amide, *i.e.* transamidation, has been reported only once and involved a highly activated N-2,2,2-trifluoroethyl amide as donor [8].

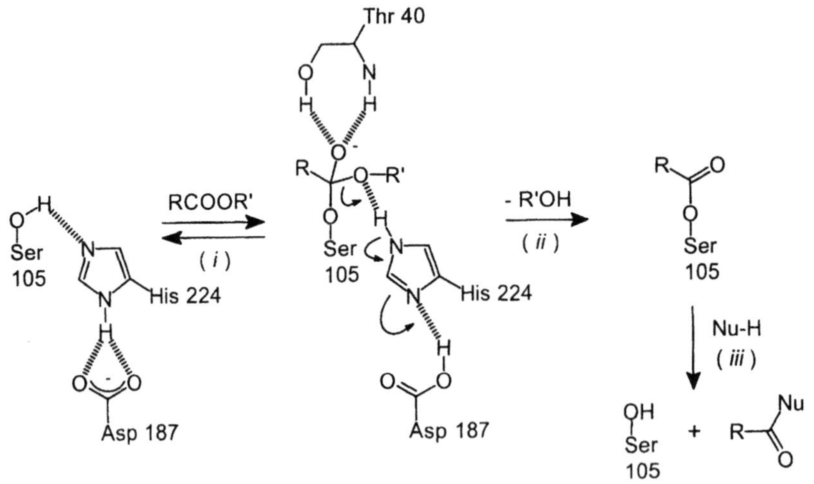

Fig. 1. Reaction mechanism of lipase catalysis; the numbering is for *Candida antarctica* lipase, - - - - - denotes a hydrogen bond; step (*iii*) is the microscopic reversal of steps (*i*) and (*ii*)

Although lipases have not been designed for enantioselectivity, they are chiral and, hence, inherently enantioselective towards all components of the reaction: the acyl group, the leaving group, and the nucleophile. The resolution of chiral amines is one of the major practical applications of lipase-catalyzed aminolysis.

The Use of Lipases in Aminolysis

Lipases[1] are ubiquitous in Nature. Initially the very impure [9] porcine pancreas lipase (PPL) – easily obtained from slaughter waste – and *Candida rugosa*[c2] lipase (CrL), a rather unstable enzyme actually consisting of a mixture of isozymes, were widely used. The recombinant microbial lipases that are now widely available generally do not contain contaminating activity and are much more stable. Some of these have been developed for application in synthetic biotransformations, but most of them were originally intended to be used in laundry formulations or the upgrading of food triglycerides.

Water, which is the natural reaction medium of enzymes, is not a good solvent for aminolysis, because hydrolysis is bound to predominate. In consequence, organic solvents are used almost universally. Even trace quantities of water in the reaction mixture will lead to undesirable hydrolysis of the acyl donor; apart from the resulting loss in yield, the liberated acid may effect deactivation of the lipase. Hence, aminolysis is preferably carried out in a strictly anhydrous medium; in some cases, activated zeolites have been added to the reaction mixture to ensure anhydrous reaction conditions. Lipases, in particular those from microbial sources that are commonly used nowadays, are exceptionally rugged enzymes and tolerate anhydrous organic solvents rather well. In particular, *Candida antarctica* lipase B (CaLB) [10, 11], which actually seems to prefer anhydrous conditions [12], has been widely used in consequence. We note that some lipases, such as that from *Thermomyces lanuginosus*[3] (TlL), that require some water to catalyse esterification, still act as an ammoniolysis catalyst in dry medium [13]. Presumably, ammonia acts as a water replacement.

We have found that ammoniolysis catalyzed by Novozym 435 (CaLB on Lewatit E) is sometimes accompanied by unexpected hydrolysis of the reactant in spite of rigorous exclusion of moisture. Apparently, the acrylic carrier strongly adsorbs traces of water that are flushed out by ammonia. This could be easily remedied by using the hydrophobic Accurel EP 100 (a macroporous polypropylene) as lipase carrier [7, 14].

The efficient use of enzymes in non-aqueous media necessitates their immobilization on a suitable carrier material [15], presumably because the use of 'free' protein dispersions renders the major part of the enzyme molecules inaccessible to the reaction medium. Nevertheless, suspensions of lyophilisate are still often used. Adsorption at the non-polar surface of organic polymers has evolved into a dominant technique for the immobilization of lipases. Lipases generally adsorb readily at a non-polar surface, because in their natural function

---

[1] A review of the structure, mechanism, and application of lipases is given in Ref. [3]

[2] *Candida rugosa* was previously known as *Candida cylindracea*

[3] Previously classified as *Humicola lanuginosa*

they adsorb at the water-lipid interface; a number of lipases are even interfacially activated. It should be kept in mind, however, that adsorption is a reversible process in aqueous medium [15]. Moreover, a major part of the activity is lost [15], which is probably caused by the partial disruption of the native structure by the hydrophobic interaction of lipase and carrier [16]. We have shown, nevertheless, that in ammoniolysis lipases adsorbed on macroporous polypropylene (Accurel EP100) are more efficient than a free suspension of an equal amount of lyophilisate [15].

Alternatively, lipases can be covalently bound to a solid carrier such as Eupergit C. The activated carrier is expensive, however, and its protein binding capacity is low. The encapsulation of lipases in hydrophobic sol-gel materials which has recently been developed [17] shows considerable potential. The very stable lipase cross-linked crystals [18] (CLCs) such as ChiroCLEC-CR and -PC from CrL and *Burkholderia cepacia*[4] lipase (BcL), respectively, have the advantage that the catalyst material is 100% protein. It has also become clear recently that CLCs of CrL maintain their activity under (aminolysis) conditions that deactivate the native lipase [19].

A number of lipases are commercially available as adsorbed, covalently bound, encapsulated, or cross-linked immobilisates. Moreover, adsorption of lipases on Accurel EP 100 is simple and straightforward [20, 21]. Finally, the modification of lipases to render them soluble in organic media deserves to be mentioned. Although a solubilized lipase has been employed in the first published example of lipase-catalyzed aminolysis [4], this methodology has not found wide application.

Synthesis of Carboxylic Amides

The lipase-catalyzed aminolysis of carboxylic esters has first been described in 1984 by *Inada et al.* [4] in a paper that also constituted the first use of a lipase in organic solvent, closely followed by a publication by *Zaks* and *Klibanov* [5].

Amines

Aminolysis seems to be a fairly general reaction that has been reported for a wide range of alkylamines, including allyl- and benzylamine [22]. Aniline derivatives are weak bases and – similar to phenols – weak nucleophiles. Only a few examples of lipase-catalyzed anilide synthesis, which required massive amounts of CrL, have been published [23,24]. Hydroxyalkylamines are preferentially acylated at the nitrogen atom [25,26], as would be expected on the basis of nucleophilicity as well as bond strength. The competition between the acylation of a secondary amine and a primary alcohol function was less one-sided [27, 28], and the actual outcome of the acylation of N-methylglucamine (Fig. 2) depended on the ionization state and, hence, the nucleophilicity of the nitrogen atom [28].

Although ammonia is the simplest amine, its use as acyl acceptor – ammoniolysis – was reported only fairly recently by us [29] as well as others [22]. Many lipases catalyzed this reaction, although some do not maintain their full

[4] *Burkholderia cepacia* was previously classified as *Pseudomonas capacia*

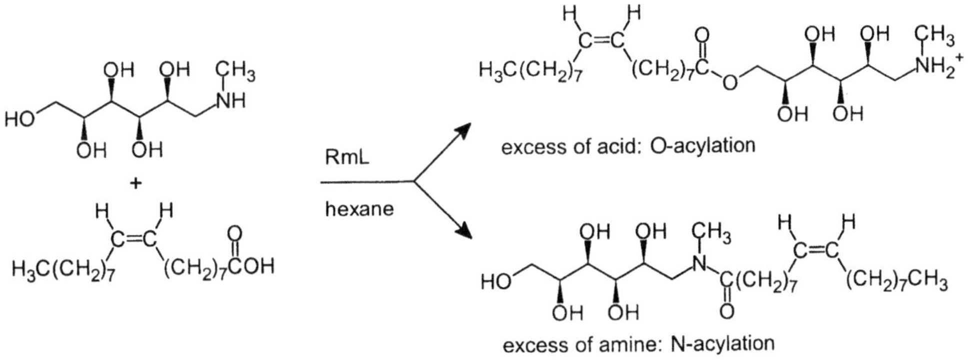

Fig. 2. O-*vs.* N-acylation of N-methylglucamine mediated by *Rhizomucor miehei* lipase (RmL)

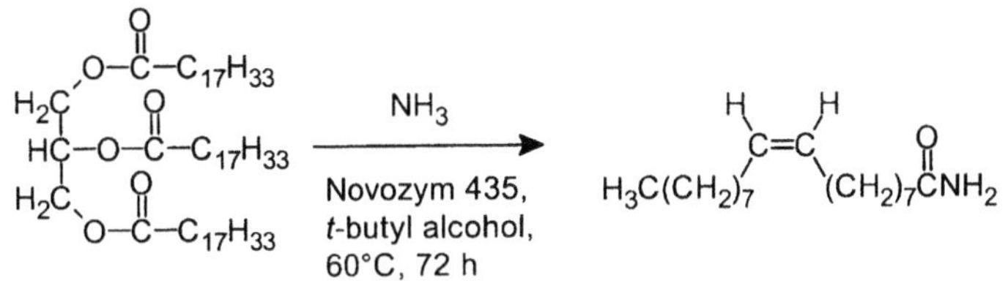

Fig. 3. Ammoniolysis of olive oil

activity at high ammonia concentrations [30]. CaLB (Novozym 435) is particularly stable and efficient in ammoniolysis as well as in aminolysis is general. It has been used, for example, in the transformation of triolein (olive oil) into oleamide (see Fig. 3) [31]. A potential advantage of the enzymatic procedure is that it affords a product of higher purity which obviates the need for purification *via* distillation.

Hydroxylamine and hydrazine, which are also weak bases but, due to the α-effect, strong nucleophiles, readily act as acyl acceptor. Accordingly, the hydroxylaminolysis of fatty acid esters, catalyzed by *Rhizomucor miehei* lipase (RmL), afforded the corresponding hydroxamic acids (N-hydroxyamides) [32]. Hydrazinolysis was, remarkably, first reported for a number of – deactivated – acylhydrazines [33] and only later on also for hydrazine proper [34], using massive amounts of Amano P lipase in both cases.

The hydroxylaminolyis and hydrazinolysis of ethyl octanoate have been extensively investigated in our group [20]. Both nucleophiles require the presence of at least one equivalent of water to prevent irreversible deactivation of the biocatalyst. Hence, hydroxylaminolysis is accompanied by partial hydrolysis of the acyl donor, whereas in hydrazinolysis the hydrolytic side reaction was almost absent. Because the acid resulting from hydrolysis also was converted [32a], although at a lower rate (see below), a nearly quantitative conversion into the hydroxamic acid was eventually achieved.

Acyl donors: carboxylic esters

A wide range of esters, including α,β-alkenic [35, 36] and β-keto esters [22] have been used as acyl donor. Ethyl propiolate, which gave *Michael* adducts with

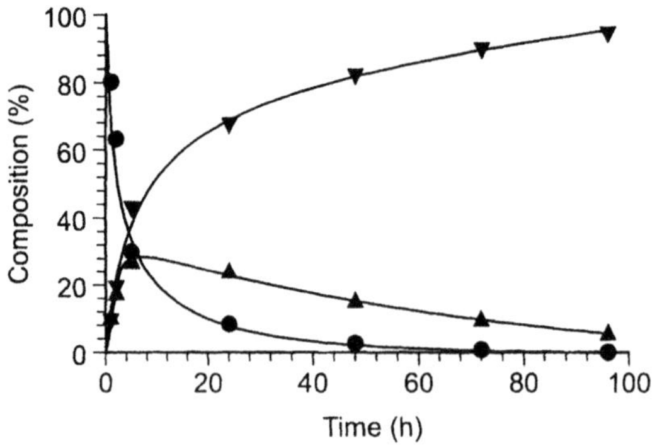

Fig. 4. Time course of the hydroxylaminolysis and competing hydrolysis of ethyl octanoate catalyzed by TIL [20]; ●: ethyl octanoate, ▲: octanoic acid, ▼: octanohydroxamic acid; reaction conditions: 688 mg (4 mmol) ethyl octanoate, 0.5 *M* NH_2OH, 1.75 *M* H_2O, 10 mg immobilized TIL on Accurel EP 100 in 10 cm^3 *tert*-butyl alcohol, 40°C

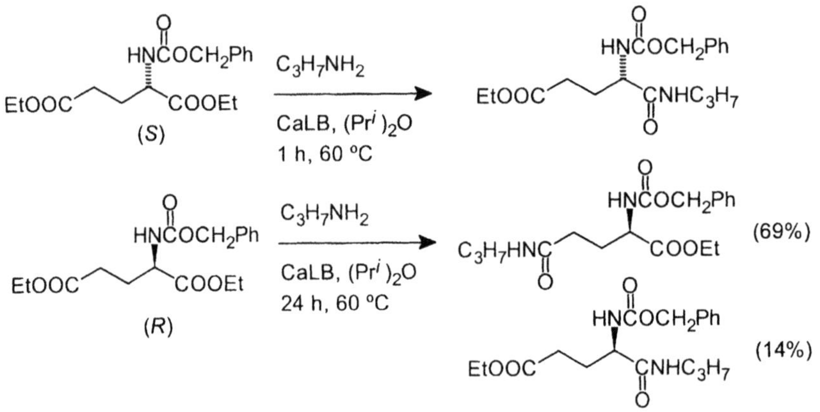

Fig. 5. Aminolysis of diethyl (*S*)- and (*R*)-N-benzyloxycarbonylglutamate

alkylamines [23], reacted with aniline derivatives and ammonia [35] to give the corresponding amides in high yield. The selective transformation of diesters to the monoamide has been observed in hydrophilic solvents [36–39], whereas formation of the succinimide was observed upon aminolysis of diethyl succinate in hexane [38]. An interesting effect of the absolute configuration of the reactant on the regioselectivity is presented by diethyl N-benzyloxycarbonylglutamate (Fig. 5) [37].

Lactams were readily formed by the intramolecular aminolysis of 4- and 5-amino-alkanoic esters, but 6-aminoalkanoic esters reacted only sluggishly to give the ε-lactams [40].

Dialkyl carbonates, the close structural relatives of carboxylic esters, have not often been subjected to lipase-catalyzed transformations. We have found that only CaLB converted carbonates from dibutyl carbonate upwards at a useful rate [41]. Reaction with alkyl amines or ammonia afforded the corresponding carbamate without any further reaction to the ureum derivative (Fig. 6).

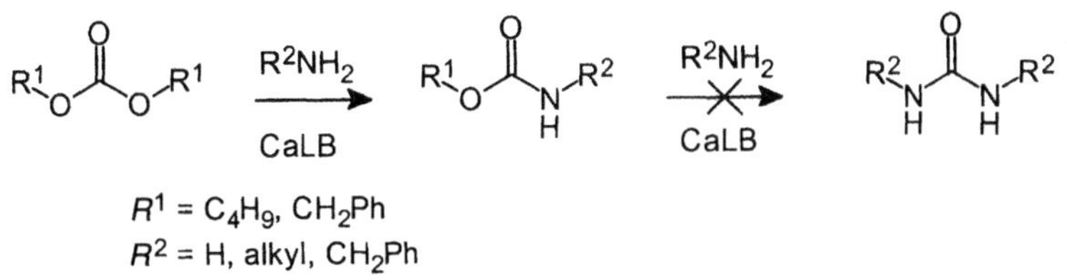

Fig. 6. Aminolysis of dialkyl carbonates

Peptide synthesis

Lipase-catalyzed peptide synthesis offers a number of potential advantages compared with competing methodologies. In contrast to chemical synthesis, circuitous protection/deprotection strategies are not required. Compared with proteases, which are restricted to *L*-amino acids, lipases offer a much more relaxed substrate specificity; moreover, product hydrolysis is virtually absent. These advantages were recognized quite early, and a number of lipase-mediated dipeptide syntheses have been published from 1987 onwards [42–45]. In order to impart the desired selectivity, the amino function in the acyl donor was protected with an acyl group. Out of a large number of lipases tested, only PPL mediated the synthesis at a useful rate [45]. PPL was specific for donors with *L*-configuration [44], whereas the enantiomeric preference towards the acceptor was generally slight [42, 45] (Table 1). These results should be interpreted with some caution, because crude PPL is known to contain protease and carboxylesterase contaminants which, in principle, could be the origin of the observed activity [43]. CrL, which was much less active than PPL, preferentially converted donors with *D*-configuration [44].

Table 1. PPL mediated dipeptide synthesis

Donor	Acceptor	Solvent	Product	Time (h)	Yield (%)	Ref.
Ac-N-*L*-Phe-OEtCl	*L*-Leu-NH_2	Toluene	N-*Ac*-*L*-Phe-*L*-Leu-NH_2	72	83	[42]
Ac-N-*L*-Phe-OEtCl	*D*-Leu-NH_2	Toluene	N-*Ac*-*L*-Phe-*D*-Leu-NH_2	>72	76	[42]
Z-*L*-Phe-OMe	*L*-Ala-OBu^s	Et_2O	*Z*-*L*-Phe-*L*-Ala-OBu^s	16	85	[43]
Z-*L*-Phe-OMe	*D*-Ala-OBu^s	Et_2O	*Z*-*L*-Phe-*D*-Ala-OBu^s	16	60	[43]
Bz-N-*L*-Tyr-OEt	*L*-Thr-OMe	3-Me-pentanol	*Bz*-N-*L*-Tyr-*L*-Thr-OMe	7	92	[45]
Bz-N-*L*-Tyr-OEt	*D*-Thr-OMe	3-Me-pentanol	*Bz*-N-*L*-Tyr-*D*-Thr-OMe	7	73	[45]

Acyl donors: carboxylic acids

Because carboxylic acids and amines tend to form unreactive salts, it was generally assumed that their condensation to the corresponding amides would not be feasible. This opinion has persisted until recently in spite of a number of contrary results that have appeared in the literature. *Tuccio et al.* found that a number of lipases catalyzed the condensation of alkylamines and fatty acids – ranging from acetic to palmitic acid – in anhydrous hexane [46]. The condensation of taurine and oleic acid (Fig. 7) was mediated by Rml as well as CaLB [47], and it has been shown

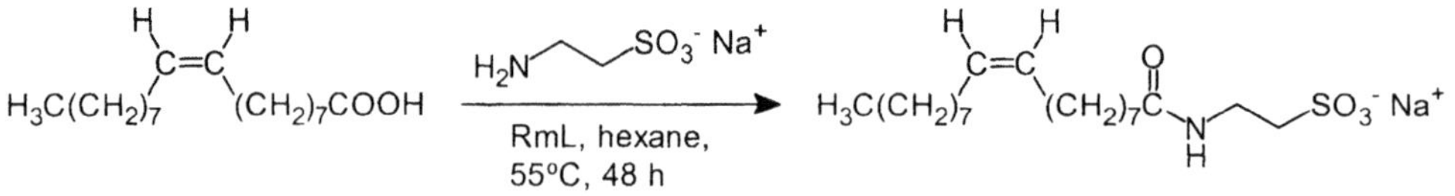

Oleic acid/taurine (m*M*/m*M*)	Acid conversion (%)
120/30	0
120/60	6
120/120	25
120/240	18
120/480	0

Fig. 7. Condensation of taurine and oleic acid

that for optimum yield the acid and amine reactants should be present in equal amounts to maximize the concentration of both reactants in solution. The synthesis of N-oleyl-N-methylglucamine from oleic acid and N-methylglucamine has been investigated by the same group [28]. Quite recently, *Litjens et al.* showed that the lipase-catalyzed reaction of butyric acid and ammonia is quite feasible [48]; these authors have also established the parameters of the equilibria involved [49].

The lipase-catalyzed condensation of carboxylic acids with hydroxylamine and hydrazine [20, 32a, 50] as well as their derivatives [20] presents a special case because of the stability of the resulting amide bond. These reactions proceed readily to essentially quantitative conversions (Table 2); the condensation of octanoic acid and hydroxylamine has even be performed in water [20]. The advantage of starting from the acid instead of the ester is twofold: it may save a reaction step, and there is virtually no uncatalyzed background reaction, which is important in kinetic resolutions (see later).

Even more surprisingly, many lipases also mediated the condensation of a carboxylic acid with an acyl hydrazide to the N,N′-diacylhydrazine (Fig. 8) [51].

Table 2. Acylation of hydroxylamine and hydrazine derivatives

Donor	Ethyl octanoate	Octanoic acid	
Nucleophile	Initial rate ($\mu mol \cdot g^{-1} \cdot min^{-1}$}	Initial rate ($\mu mol \cdot g^{-1} \cdot min^{-1}$)	Yield[a] (%)
H_2NOH	910	160	99
H_2NOPh	n.d.	<5	66
H_2NOCH_2Ph	200	87	98
$HN(CH_2Ph)OH$	n.d.	<5	65
H_2NNH_2	540	70	91
H_2NNHPh	52	70	95
H_2NNHCH_2Ph	200	100	96

[a] Reaction conditions: 57.6 mg (0.4 mmol) octanoic acid, 1.0 mmol nucleophile, and 50 mg Novozym 435 in 1.0 cm^3 *tert*-butyl alcohol, 20 h, 40°C

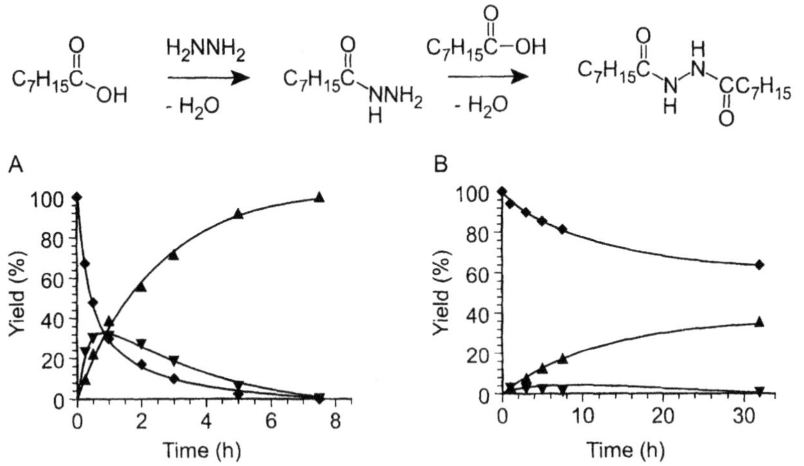

Fig. 8. Time course of the reaction of hydrazine and octanoic acid in isooctane [51]; lipase: A: TIL, B: PaL; ◆: octanoic acid, ▼: octanohydrazine, ▲: N,N′-dioctanoylhydrazine; reaction conditions: 1 mmol hydrazine hydrate, 2 mmol octanoic acid, 50 mg immobilized lipase in 5 cm^3 isooctane, 40°C; 100% yield of N,N′-dioctanoylhydrazine equals 1 mmol

TIL was conspicuous for its high activity in this reaction. Some lipases, such as *Pseudomonas* lipoprotein lipase, *P. alcaligenes* lipase (PaL), and *C. antarctica* lipase A acylated octanohydrazide faster than hydrazine itself when both nucleophiles were present in the reaction mixture.

Enantioselective Aminolysis

Although lipases have not been designed for enantioselectivity, it was recognized quite early [52] that they are chiral and, hence, inherently enantioselective. This feature has been exploited in aminolysis with the purpose of resolving chiral acyl donors as well as amines. An inherent advantage of aminolysis as a resolution strategy is that, due to the stability of the amide bond, the formation of a carboxylic amide is virtually irreversible [7], which is a prerequisite for an efficient resolution [53].

It is common practice to discuss a kinetic resolution in terms of its enantiomeric ratio E [54] which is equal to the ratio of the *pseudo*-first-order rate constants of the enantiomers according to Eq. 1. We note that formally this practice is not correct for lipase-catalyzed resolutions, because lipases do not obey minimal *Michaelis-Menten* kinetics and, hence, slight deviations of the predicted behaviour are to be expected [55].

$$E = (k_{cat}/K_m)_R/(K_{cat}/K_m)_S \quad (1)$$

Resolution of the acyl donor

Because of the ping-pong mechanism of lipases the conversion of the acyl donor involves a number of steps (see Fig. 1), each of which is subject to – potentially conflicting – enantiomer discriminating effects. The resulting enantiomeric

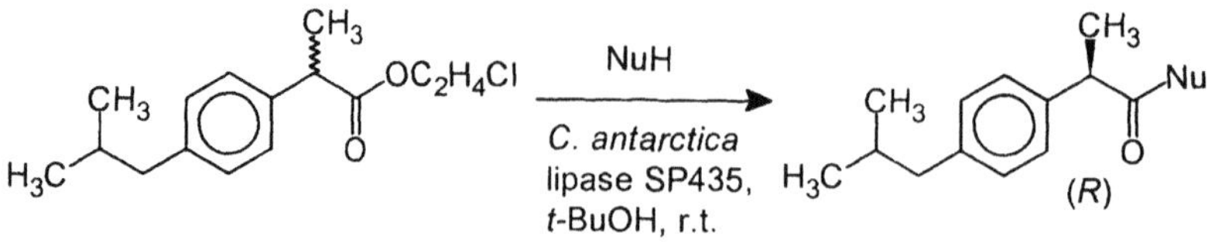

NuH	Conv. (%)	ee_{ester} (%)	*E*
H_2O	63	5.8	4
BuOH	56	86	16
NH_3	56	96	28

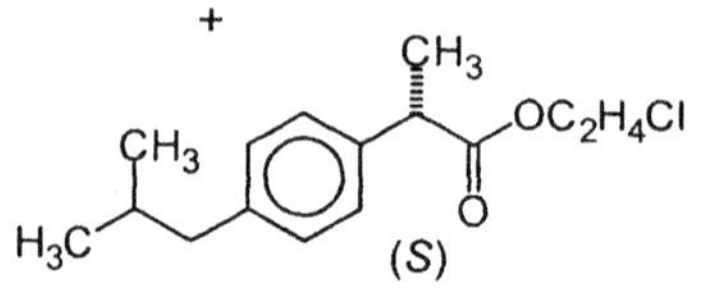

Fig. 9. Effect of the acceptor on the CaLB-catalyzed resolution of ibuprofen

preference of the transformation is the sum of all these enantiodiscriminatory processes. Hence, the outcome may depend on the reaction conditions and, in particular, on the nature of the acyl acceptor. The CrL mediated aminolysis of methyl 2-chlorobutanoate, for example, was moderately (*S*)-selective ($E = 9$)[5] [8, 56, 57]. In contrast, the esterification of 2-chloropropionic acid, mediated by the same catalyst, was highly (*R*)-specific [58, 59], whereas the hydrolysis of methyl 2-chlorobutanoate was only slightly biased in favour of the (*R*)-enantiomer ($E = 2$) [57, 60]. The N-alkyl-azetidine-2-carboxylates (**1**, Fig. 12) present a comparable case: their resolution could not be accomplished by enzymatic hydrolysis [61], but ammoniolysis catalyzed by CaLB gave the (*R*)-ester and the (*S*)-amide with high selectivity [62]. A final example is the CaLB mediated ammoniolysis of the chloroethyl ester of ibuprofen, which was much more enantioselective than its hydrolysis or alcoholysis (Fig. 9) [29, 30].

In contrast with CaL, CrL is known to convert the physiologically active (*S*)-ibuprofen preferentially [63]. This feature was exploited in the synthesis of (*S*)-ibuproxam, a prodrug, by condensation of racemic ibuprofen with hydroxylamine in aqueous medium (Fig. 10). Crosslinked crystals of CrL (CLEC-CR) were used as the catalyst, because the native enzyme was deactivated under the reaction conditions [19].

Acyl donor resolution would become much more interesting if it could be combined with racemization of the slow-reacting enantiomer to achieve a 100%

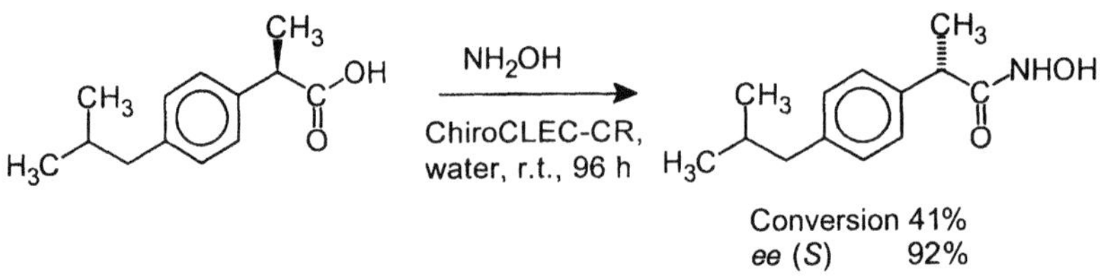

Fig. 10. CLEC-CR mediated synthesis of (*S*)-ibuproxam

[5] *E*-factors, when not provided by the authors, were calculated from the conversion (c) and the enantiomeric purity of the product (ee_p) using the equation derived by *Chen et al.* [54]: $E = \ln(1 - c \cdot (1 + ee_p))/\ln(1 - c \cdot (1 - ee_p))$

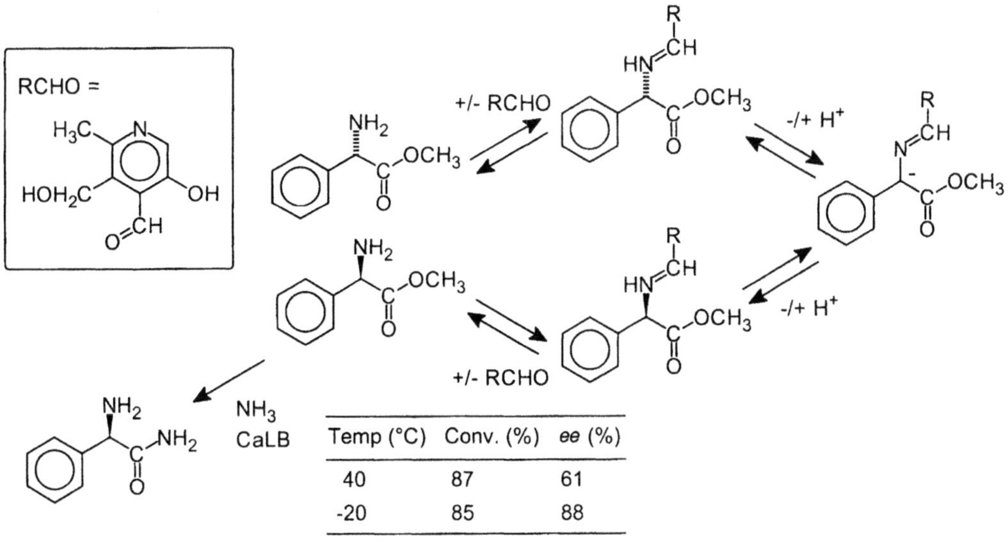

Temp (°C)	Conv. (%)	*ee* (%)
40	87	61
-20	85	88

Fig. 11. Ammoniolytic synthesis of (*R*)-phenylglycine amide by dynamic kinetic resolution

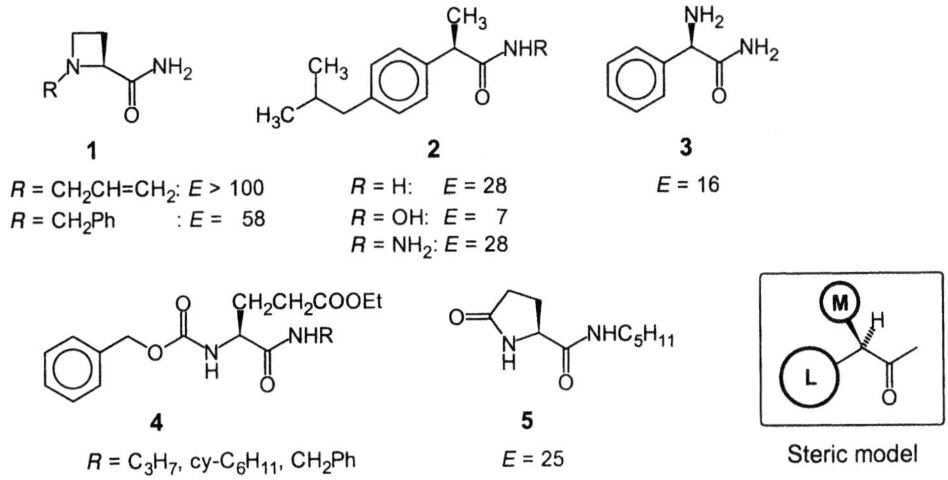

Fig. 12. Enantiorecognition of chiral acyl groups by CaLB

yield of enantiomerically pure product. This has been demonstrated for the CaLB mediated synthesis of (*R*)-phenylglycine amide from its racemic ester [64]. By combining the ammoniolysis with *in situ*-racemization of the slow-reacting (*S*)-ester *via* a *Schiff* base, a dynamic kinetic resolution was developed [65]. Initially the *ee* of the amide was rather low, but it could be improved significantly by lowering the reaction temperature (Fig. 11) [14].

The enantiomeric preference of CaLB in the aminolysis of α-substituted acyl donors is summarized in Fig. 12. The preferentially formed enantiomers of **1** [62], ibuprofen amides (**2**) [19, 29, 30], phenylglycine amide (**3**) [64], and derivatives of glutamic acid (**4**) [37] seem to correspond with a common steric model with the substituents arranged as indicated in Fig. 12. Only the selective ($E = 25$) formation of (*S*)-N-pentylpyroglutamic amide (**5**) [66] does not easily fit into the pattern.

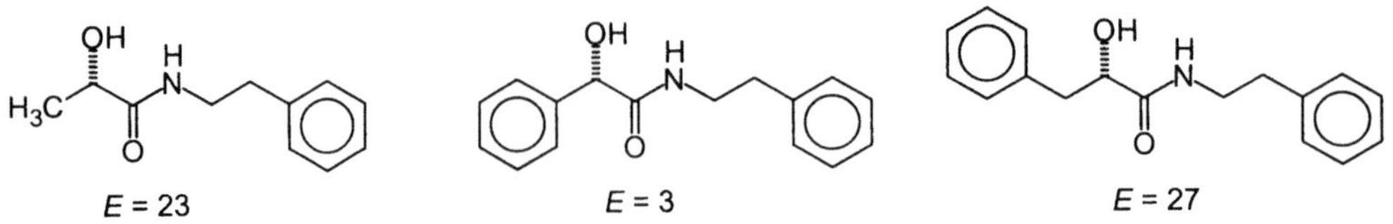

Fig. 13. BcL mediated resolution of benzyl α-hydroxycarboxylates

A number of α-hydroxycarboxylic acid benzyl esters were transformed into their benzylamides in the presence of BcL [67]. The (*S*)-amides were preferentially formed (Fig. 13), whereas the CaLB mediated aminolysis of α-aminocarboxylic acid esters was (*R*)-specific (Fig. 12).

Chiral 3-substituted acyl groups are hardly less efficiently discriminated by CaLB than 2-subsituted ones, in spite of the increasing distance from the reactive centre. The ethyl ester of 3-aminobutyric acid [68] was converted into benzamide **6** with moderate enantioselectivity, but the corresponding 3-hydroxy compound (**7**) was formed with almost absolute enantioselectivity [69, 70] (Fig. 14); the ammoniolysis of ethyl 4-chloro-3-hydroxybutyrate likewise was highly selective [71]. In contrast, BcL acted with negligible enantioselectivity in the benzylaminolysis of ethyl 3-hydroxybutyrate [69].

Prochiral diesters, such as dimethyl 3-hydroxyglutarate, are interesting reactants because their monoaminolysis could, in principle, afford a 100% yield of enantiopure product. Indeed it was found [72] that the *pro-R* ester group in dimethyl 3-hydroxyglutarate was preferentially converted to the monoamides **9** with complete selectivity by CaLB mediated aminolysis (Fig. 14). The enantioselective aminolysis of chiral vinyl carbonates to the corresponding carbamates (**10**), which is not really an acyl resolution, was also quite efficient [73].

It has been explained above that the enantiorecognition of a chiral acyl group involves many steps. Hence, the structure of the amine would be expected to contribute to the overall effect, and intuitively one would expect enantiorecognition to become more efficient with sterically demanding amines. Systematic studies in this field have been few; the available data, which involve β-amino- and β-hydroxyesters, are compiled in Table 3. Based on this very restricted data set, benzylamine would seem to be the resolving agent of choice.

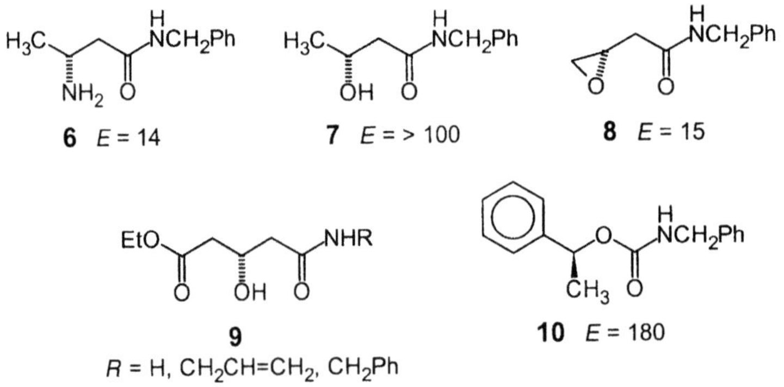

Fig. 14. Enantiomeric preferences of CaLB in the aminolysis of β-substituted acyl donors

Table 3. Effect of the amine on the enantiorecognition of CaLB[a]

Product	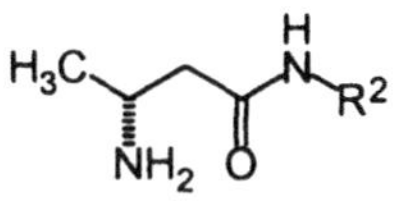	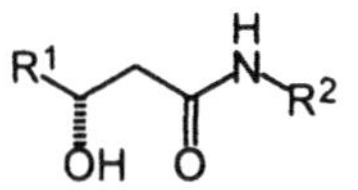			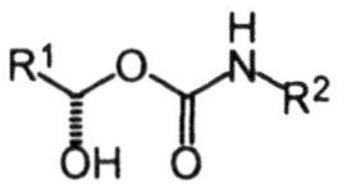		
R^1		CH_3	CH_2CH_3	CH_2CH_2Cl	CH_2CH_3	n-C_6H_{13}	C_6H_5
$R^2 = H$	4		44[b]				
R^2 = n-Bu		17	27	43	90	24	63
R^2 = Benzyl	14	> 100	52	> 100	700	200	183
R^2 = n-Octyl		30	> 100	20	160	10	22
Reference	68	70	70	70	73	73	73

[a] Enantiomeric ratios as taken from the references or calculated from conversion and *ee*; [b]data taken from Ref. [71]

Solvents often exert dramatic effects on enzymatic resolutions [74], and an inversion of the enantiopreference has even been observed in certain cases. The effect of the solvent on the CaLB mediated condensation of ibuprofen and hydroxylamine to (*R*)-ibuproxam (**2**, $R = OH$) was only modest, however [19]. E ranged from a value of 4 in water and some moderately hydrophilic solvents *via* 8 in isooctane or *tert*-butyl alcohol to 13 in dioxane. The enantiomer discrimination in the CaLB-catalyzed ammoniolysis of phenylglycine methyl ester (Fig. 11) was likewise little affected by the nature of the solvent [14].

Resolution of chiral amines

Acylation studies of secondary alcohols have shown that the nucleophile subsites of all lipases have the same steric preference. A model that has originally been constructed for CrL and BcL [75] was later found to be valid for many other lipases – although not always to the same extent [76] – including CaLB [77]. An overview of the aliphatic and arylaliphatic amines that have been resolved by lipase-catalyzed acylation is given in Fig. 15. The configuration of the preferentially converted enantiomer corresponds with the steric model mentioned above.

The magnitude of the enantiomeric preference, *i.e.* E, depends, often critically, on the nature of the acyl reagent and the solvent. Systematic studies in this field are almost non-existent, but we have compiled the available data, which mainly involve CaLB, in Table 4. From these it becomes clear that almost any secondary alkylamine can be resolved once the right reaction conditions have been found. In general, a bulky acyl group, which restricts the conformational freedom of the nucleophile, would be expected to assist the enantiodiscrimination. Indeed 2-aminobutane (**11a**) was much more efficiently resolved by methyl methacrylate than by methyl acrylate [23b]. However, dimethyl succinate also worked rather well in dioxane; in hexane the enantioselectivity was poor [38]. This latter pattern, with the best results being obtained in mildly hydrophilic media, will often be repeated although it is by no means universal.

2-Aminopentane (**11b**) has been resolved, also using CaLB, under completely different conditions: by ethyl acetate in a solvent-free medium [78]. A comparable

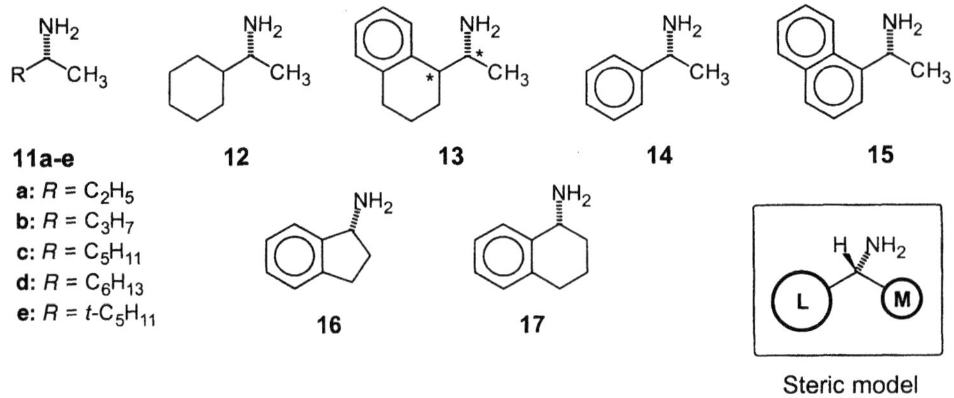

Fig. 15. Preferentially acylated enantiomers of aliphatic and arylaliphatic amines[6]

result was obtained with *Pseudomonas aeruginosa* lipase (PaL) in diethyl ether [79]. The bulky dibenzyl carbonate was an unexpectedly inefficient resolving agent [41], perhaps because hexane is not the optimum solvent considering the solvent effect in the acylation of **11a** [38].

In the resolution of 2-aminoheptane (**11c**) *via* acylation by dimethyl succinate the solvent effect was the opposite of that observed earlier: a good enantiomeric ratio ($E = 68$) was obtained in hexane, whereas reaction in dioxane resulted in only half that value [38]. Another attempt to resolve **11c** in dioxane using the activated reagent octyl vinyl carbonate also failed [80]; **11a** was not resolved by this system either. Finally, ethyl octanoate, an acyl donor that has not been mentioned so far, resolved 2-aminooctane (**11d**) almost quantitatively [81] in a solventless and, hence, hydrophobic reaction medium. However, dibenzyl carbonate in hexane again was quite inefficient [41].

The branched and, therefore, more sterically restricted 2-amino-3,3-dimethylbutane (**11e**) was acylated with low enantioselectivity by neat ethyl acetate [82], whereas this reaction system had given excellent results with **11b** [78]. Even more surprisingly, the *E* factor of **11e** improved by more than an order of magnitude when the acyl donor was isopropyl acetate. Reaction in 1,2-dimethoxyethane (*DME*), which is a more hydrophilic medium, resulted in a further improvement by a factor of seven[6].

1-Cyclohexylethylamine (**12**) was acylated with a moderate *E* of 17 by dibenzyl carbonate in toluene [83]. The more hindered and less flexible amine **13** behaved similarly to **11e** upon acylation by acetic acid esters in the presence of CaLB: the enantiodiscrimination was modest when the ethyl ester was used as donor but, quite remarkably, absolute with the isopropyl ester [82]; BcL was much less selective, although the same trend became apparent.[6] It should be noted that **13** has two chiral centres; the acylation yields [82] indicate that only one (presumable that attached to the nitrogen atom) was resolved.

[6] The absolute configurations of the acylation products of **11e** and **13** were not stated by the authors [82]; by analogy with comparable resolutions, the (*R*)-enantiomer should be expected to be the fast reacting one

Table 4. Resolution of chiral amines **11–17**

Amine	Lipase	Donor	Solvent	*T*/°C	*E*	Ref.
11a	CaLB	Me acrylate	*THF*	30	8	[23b]
	CaLB	Me methacrylate	*THF*	30	55	[23b]
	CaLB	Dimethyl succinate	Dioxane	30	35	[38]
	CaLB	Dimethyl succinate	Hexane	30	6	[38]
	CaLB	Octyl vinyl carbonate	Dioxane		7	[80]
11b	CaLB	Ethyl acetate	–		> 200	[78]
	CaLB	Dibenzyl carbonate	Hexane	r.t.	4	[41]
	PaL	Ethyl acetate	Diethyl ether	r.t.	> 100	[79]
11c	CaLB	Dimethyl succinate	Dioxane	30	37	[38]
	CaLB	Dimethyl succinate	Hexane	30	68	[38]
	CaLB	Octyl vinyl carbonate	Dioxane	30	3	[80]
11d	CaLB	Ethyl octanoate	–	39	> 100	[81]
		Dibenzyl carbonate	Hexane	r.t.	5	[41]
11e	CaLB	Ethyl acetate	–	28	8	[82]
	CaLB	Ethyl acetate	*DME*	r.t.	29	[82]
	CaLB	*i*-Propyl acetate	–	28	109	[82]
	CaLB	*i*-Propyl acetate	*DME*	r.t.	> 700	[82]
12	CaLB	Dibenzyl carbonate	Toluene	r.t.	17	[83]
13	CaLB	Ethyl acetate	–	28	14	[82]
	CaLB	*i*-Propyl acetate	–	28	450	[82]
	BcL	Ethyl acetate	–	r.t.	2	[82]
	BcL	*i*-Propyl acetate	–	r.t.	28	[82]
14	CaLB	Ethyl acetate	–		110	[78]
	CaLB	Ethyl octanoate	–	39	> 200	[81]
	CaLB	Dimethyl succinate	Dioxane	30	95	[38]
	CaLB	Dimethyl succinate	Hexane	30	6	[38]
	CaLB	Octyl vinyl carbonate	Dioxane		10	[80]
	CaLB	Octyl vinyl carbonate	Di-*i*-propyl ether		200	[80]
	CaLB	Octyl vinyl carbonate	Hexane		200	[80]
	CaLB	Dibenzyl carbonate	Toluence	r.t.	> 300	[83]
	CaLB	Dibenzyl carbonate	Hexane	r.t.	20	[83]
	CaLB	Dibenzyl carbonate	Hexane	r.t.	50	[41]
	BpL	Ethyl methoxyacetate	*MTBE*		114	[84]
15	CaLB	Ethyl acetate	–		24	[78]
	CaLB	Ethyl acetate	–	28	15	[82]
	CaLB	*i*-Propyl acetate	–	28	100	[82]
	CaLB	*i*-Propyl acetate	*DME*	r.t.	650	[82]
	BcL	Ethyl acetate	–	r.t.	3	[82]
	BcL	*i*-Propyl acetate	–	r.t.	18	[82]
16	CaLB	Dibut-3-enyl carbonate	Toluene	r.t.	> 200	[83]
	CaLB	Dibenzyl carbonate	Hexane	r.t.	72	[41]
17	CaLB	Dibenzyl carbonate	Hexane	r.t.	19	[41]

1-Phenylethylamine (**14**), which has attracted a major share of the efforts in this field, has successfully been resolved under a wide range of reaction conditions. Neat ethyl acetate [78] or octanoate [81] gave a high selectivity; a comparable result was achieved with dimethyl succinate in dioxane, but in hexane the enantioselectivity was low [38] as observed with **11a**. Octyl vinyl carbonate showed the opposite solvent effect: **14** was resolved with absolute selectivity in diisopropyl ether or hexane, but not in dioxane [80]. Dibenzyl carbonate acted with excellent selectivity in toluene [83], but in hexane [41, 83] the result was much more modest, seemingly confirming that hexane is not the optimum medium for this acyl donor.

Finally, the acyl donor methoxyacetic acid ethyl ester deserves to be mentioned. This compound was developed by BASF as a mildly activated and easily recyclable reagent for the enzymatic resolution of chiral alcohols and amines. **14** was efficiently resolved by *Bukholderia plantarii* lipase (BpL) mediated methoxyacetylation in *tert*-butyl methyl ether (*TBME*) [84].

The enantiodiscrimination of 1-(1-naphthyl)ethylamine (**15**) in the CaLB mediated acylation by neat ethyl acetate was only modest [78, 82], but E could be improved dramatically to over 600 by using isopropyl acetate in *DME* [82] similar to **11e** and **13**.

Finally, the bicyclic amine 1-aminoindane (**16**) was efficiently resolved [41, 83], but the closely related 1-aminotetraline (**17**) reacted with a moderate E of 19 [41]. A dramatic effect of the leaving group (ethyl *vs.* isopropyl) has manifested itself in the acylation of **11e**, **13**, and **15** [82]. The obvious conclusion is that the liberated alcohol influences the enantiorecognition of the amine, presumably because it binds in the active site of the lipase. It is pertinent to note that the interaction with the inihibition of lipases by hydroxylic reaction media is well documented [7, 85].

It has already been noted that *in situ*-racemization of the slowly reacting enantiomer has the potential advantage that the fast reacting enantiomer can be isolated in 100% yield. This has been accomplished with **14** by combining its CaLB mediated acylation with palladium-catalyzed reacemization of the unreacted amine in one pot (Fig. 16) [86].

A few resolutions of more bulky α-phenyl substituted amines have been carried out. The CaLB mediated resolution of a range of 1-phenyl-2-propynylamines (**18**) has been accomplished with absolute enantioselectivity in neat ethyl acetate [87]. Apparently the acetylene moiety is readily accommodated in the part of the nucleophile subsite where the M-group binds in the resolutions of **11–15**. 1-Amino-1-phenylacetonitrile (**19**), the nitrile analogue of **18**, reacted with 2,2,2-trifluoroethyl acetate in diisopropyl ether with moderate (S)-selectivity ($E = 12$–15) in the presence of PaL or *Chromobacterium viscosum* lipase (CvL), but BcL converted (R)-**19** slightly faster [88]. The CaLB mediated reaction of **19** and phenylacetic

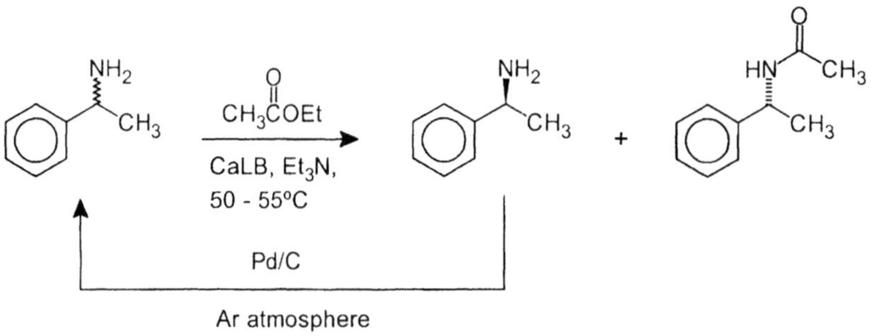

Fig. 16. Dynamic kinetic resolution of 1-phenylethylamine (**14**)

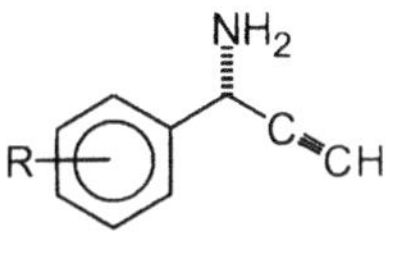

18 *E* > 100

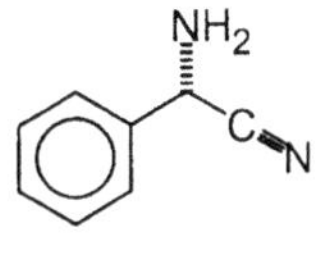

19

R = H, 4-Cl, 4-F, 3-F, 3-Me, 2-Me

Fig. 17. Enantiopreference of CaLB in the acylation of **18** and **19**

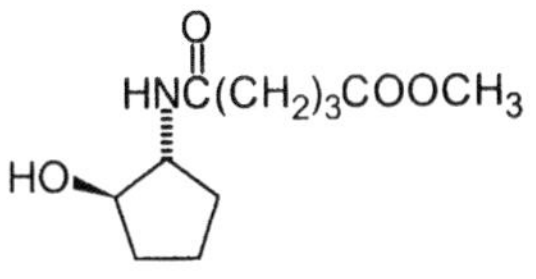

20 *E* = 4

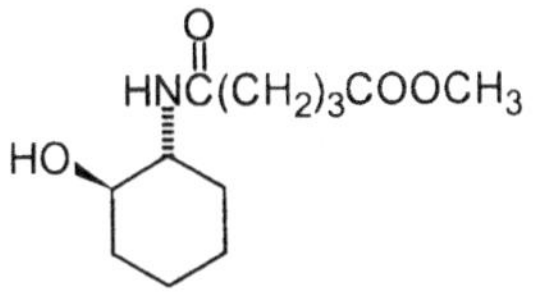

21a *E* = 31

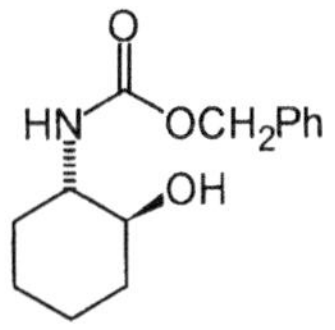

21b *E* = 16 - 25

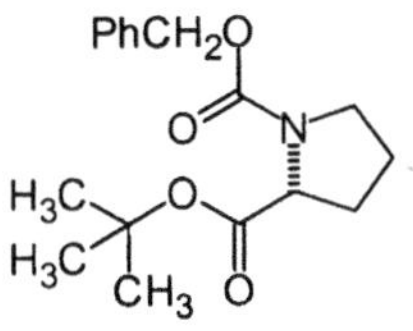

22 *E* = 4

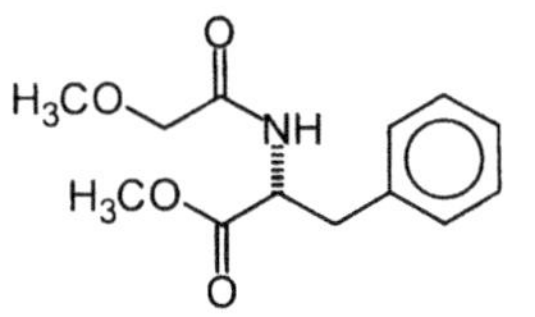

23 *E* >> 100

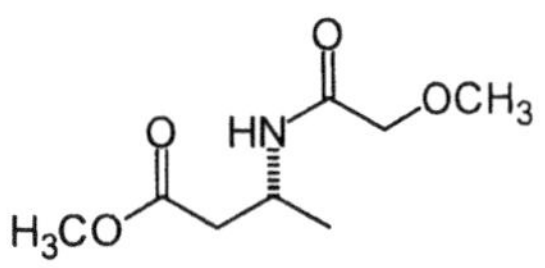

24 *E* >> 100

Fig. 18. CaLB mediated enantioselective acylation of functionalized amines **20–24**

acid methyl ester was highly (*S*)-selective [89], which may be explained as a beneficial effect of steric congestion in the active site.

A few functionalized amines, such as β-hydroxyamines and amino acid esters have been resolved by enantioselective acylation (Fig. 18), and the question arises whether or not the same steric model applies. The CaLB mediated acylation of *trans*-1-amino-2-hydroxycyclopentane with dimethyl glutarate to **20** was only modestly enantioselective [90]. The enantiomers of its C_6-homologue were more readily discriminated (**21a**) when the reaction was carried out in dioxane, but no resolution was possible in toluene. This enantiopreference would seem to be consistent with the steric model, with the hydroxyl group binding as the *L* substituent. Surprisingly, the enantiopreference was inverted with dibenzyl carbonate as the acyl donor [83, 90]; **21b** was formed with $E = 16$ in either dioxane or toluene. The acylation of amino acid esters was (*R*)-selective, as would be expected on the basis of the peptide synthesis experiments. Proline *tert*-butyl ester (**22**) was inefficiently resolved using dibenzyl carbonate [83] but, surprisingly, the acylation of the methyl esters of phenylalanine (**23**) and β-aminobutyric acid (**24**) with methoxyacetic acid ethyl ester resulted in a nearly quantitative ($E = 200$) resolution [91]. The reaction of β-aminobutyric acid ethyl ester with ethyl acetate was somewhat less enantioselective [68].

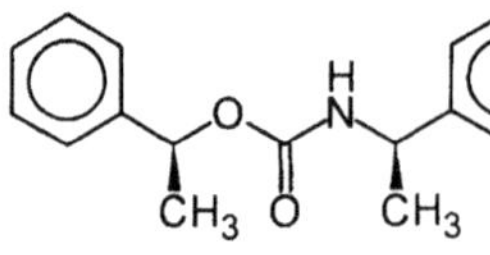

25

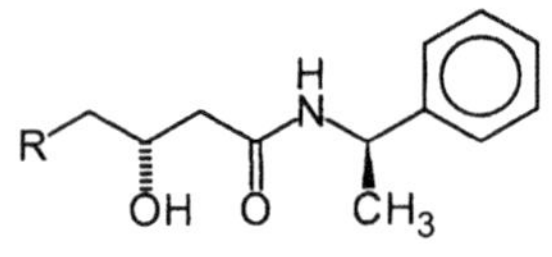

26

R = Cl, COOEt

Fig. 19. Combined resolutions

Combined resolutions

Considering that lipases can resolve the acyl donor as well as the acyl acceptor, it might be attractive to accomplish both resolutions in one reaction. Such a one-step resolution of a chiral vinyl carbonate and chiral amines, mediated by CaLB, was only moderately successful, because the products, such as *e.g.* **25** (see Fig. 19), were formed in a moderate diasteromeric and/or enantiomeric excess [92]. In contrast, **26** and some related amides were obtained as the almost pure (3*S*,1′*R*) stereoisomers [93].

Conclusion

The lipase-catalyzed aminolysis of carboxylic esters is a fairly general methodology for the synthesis of carboxylic amides under mild conditions. The condensation of carboxylic acids and the amines provides an alternative route to carboxylic amides. Lipase-catalyzed aminolysis is a useful method to resolve chiral carboxylic esters as well as chiral amines. The enantiorecognition of chiral amines often depends critically on the acylating agent and the reaction medium.

References

[1] Sheldon RA (1997) Chem Ind (London) 12; Sheldon RA (1997) J Chem Technol Biotechnol **68**: 381
[2] De Zoete MC, Van Rantwijk F, Sheldon RA (1994 Catalysis Today **22**: 563
[3] Schmidt RD, Verger R (1998) Angew Chem Int Ed Engl **37**: 1608
[4] Inada Y, Nishimura H, Takahashi K, Yoshimoto T, Saha AR, Saito Y (1984) Biochem Biophys Res Commun **122**: 845
[5] Zaks A, Klibanov AM (1985) Proc Natl Acad Sci USA **82**: 3192
[6] Pleiss J, Fischer M, Schmidt RD (1998) Chem Phys Lipid **93**: 67
[7] De Zoete MC, Kock-van Dalen AC, Van Rantwijk F, Sheldon RA (1996) J Mol Catal B: Enzym **2**: 19
[8] Gotor V, Brieva R, González C, Rebolledo F (1991) Tetrahedron **47**: 9207
[9] Faber K (1997) Biotransformations in Organic Chemistry, 3rd edn. Springer, Berlin, p 92
[10] Ishii M (1988) PCT Int Appl WO 88/02775; Heldt-Hansen HP, Ishii M, Patkar SA, Hansen TT, Eigtved P (1989) In: Whitaker JR, Sonnet PE (eds) Biocatalysis in Agricultural Biotechnology, ACS Symposium Series, vol 389. American Chemical Society, Washington, chap 11, p 158
[11] Anderson EM, Larsson KM, Kirk O (1998) Biocatal Biotransform **16**: 181
[12] De Goede ATJW, Benckhuysen W, Van Rantwijk F, Maat L, Van Bekkum H (1993) Recl Trav Chim Pays-Bas **112**: 567

[13] De Zoete MC, Kock-van Dalen AC, Van Rantwijk F, Sheldon RA (1996) J Mol Catal B: Enzym **2**: 19
[14] Wegman MA, Hacking MAPJ, Rops J, Pereia P, Van Rantwijk F, Sheldon RA (1999) Tetrahedron Asymm **10**: 1739
[15] Van Rantwijk F, Kock-van Dalen AC, Sheldon RA (1998) in: Ballasteros A, Plou FJ, Iborra JL, Halling P (eds) Stability and Stabilisation of Enzymes. Progress in Biotechnology, vol 15. Elsevier Science, p 447
[16] Haynes CA, Norde W (1994) Colloids Surf B: Biointerfaces **2**: 517
[17] a) Reetz MT, Zonta A, Simpelkamp J (1995) Angew Chem Int Ed Engl **34**: 301; b) Reetz MT, Zonta A, Simpelkamp J (1996) Biotechnol Bioeng **49**: 527
[18] St Clair NL, Navia M (1992) J Am Chem Soc **114**: 7314; Margolin AL (1996) Trends Biotechnol **14**: 223; Govardhan CP (1999) Curr Opin Biotechnol **10**: 331
[19] Hacking MAPJ, Van Rantwijk F, Sheldon RA (in prep)
[20] Hacking MAPJ, Akkus H, Van Rantwijk F, Sheldon RA (2000) Biotechnol Bioeng (in press)
[21] Pedersen S, Eigtved P (1990) PCT Int Appl WO 90/15868
[22] García MJ, Rebolledo F, Gotor V (1993) Tetrahedron Lett **34**: 6141
[23] a) Rebolledo F, Brieva R, Gotor V (1989) Tetrahedron Lett **30**: 5345; b) Puertas S, Brieva R, Rebolledo F, Gotor V (1993) Tetrahedron **49**: 4007
[24] It may be a coincidence that the only documented example of lipase catalysed synthesis of phenol esters also involved CrL; see: Abramowicz DA, Keese CR (1987) Biotechnol Bioeng **33**: 149
[25] a) Gotor V, Brieva R, Rebolledo F (1988) J Chem Soc Chem Commun 957; b) Fernández S, Brieva R, Rebolledo F, Gotor V (1992) J Chem Soc Perkin Trans 1, 2885
[26] Quirós M, Rebolledo F, Gotor V (1994) J Chem Res (M) 1917
[27] Acensio G, Andreu C, Marco JA (1991) Tetrahedron Lett **32**: 4197
[28] a) Maugard T, Remaud-Simon M, Petre D, Monsan P (1997) Tetrahedron **53**: 5185; b) Maugard T, Remaud-Simon M, Petre D, Monsan P (1997) Tetrahedron **53**: 7587; c) Maugard T, Remaud-Simon M, Monsan P (1998) J Mol Catal B: Enzym **5**: 13; d) Maugard T, Remaud-Simon M, Monsan P (1998) Biochim Biophys Acta **1387**: 177
[29] De Zoete MC, Kock-van Dalen AC, Van Rantwijk F, Sheldon RA (1993) J Chem Soc Chem Commun 1831
[30] De Zoete MC, Kock-van Dalen, AC, Van Rantwijk F, Sheldon RA (1994) Biocatalysis **10**: 307
[31] De Zoete MC, Kock-van Dalen AC, Van Rantwijk F, Sheldon RA (1996) J Mol Catal B: Enzym **1**: 109 (1996) [Erratum: **2**: 141]
[32] a) Servat F, Monted D, Pina M, Arnaud A, Ledon H, Marcou L, Graille J (1990) J Am Oil Chem Soc **67**: 646; b) Servat F, Monted D, Pina M, Arnaud A, Galzy P, Graille J, Ledon H, Marcou L (1990) Tenside Surf Det **27**: 298
[33] a) Gotor V, Astorga C, Rebolledo F (1990) Synlett 387; b) Astorga C, Rebolledo F, Gotor V (1991) Synthesis 350
[34] Astorga C, Rebolledo F, Gotor V (1993) Synthesis 287
[35] Sánchez V, Rebolledo F, Gotor V (1994) Synlett 529
[36] Quirós M, Astorga C, Rebolledo F, Gotor V (1995) Tetrahedron **51**: 7715
[37] Chamorro C, Gonzáles-Muñiz R, Conde S (1995) Tetrahedron Asymm **6**: 2343
[38] Puerta S, Rebolledo F, Gotor V (1995) Tetrahedron **51**: 1495
[39] Conde S, López-Serrano, Fierros M, Biezma MI, Martínez A, Rodríguez-Franco MI (1997) Tetrahedron **53**: 11745
[40] Gutman AL, Meyer E, Yue X, Abell C (1992) Tetrahedron Lett **33**: 3943
[41] Hacking MAPJ, Van Rantwijk F, Sheldon RA (2000) J Mol Catal B: Enzym (in press)
[42] Margolin AL, Klibanov AM (1987) J Am Chem Soc **109**: 3802
[43] West JB, Wong C-H (1987) Tetrahedron Lett **28**: 1629
[44] Kawashiro K, Kaise, K, Minato D, Sugiyama S, Hayashi H (1993) Tetrahedron **21**: 4541
[45] So J-E, Kang S-H, Kim B-G (1998) Enzyme Microb Technol **23**: 211

[46] Tuccio B, Ferré E, Comeau L (1991) Tetrahedron Lett **32**: 2763
[47] Maugard T, Remaud-Simon M, Petre D, Monsan P, (1997) Biotechnol Lett **19**: 751
[48] Litjens MJJ, Straathof AJJ, Jongejan JA, Heijnen JJ (1999) Chem Commun (Cambridge) 1255
[49] Litjens MJJ, Sha M, Straathof AJJ, Jongejan JA, Heijnen JJ (1999) Biotechnol Bioeng **65**: 347
[50] Vaysse L, Dubreucq E, Pirat J-L, Galzy P (1997) J Biotechnol **53**: 41
[51] Hacking MAPJ, Van Rantwijk F, Sheldon RA (2000) J Mol Catal B: Enzym (in press)
[52] Dakin HD (1903) Proc Chem Soc **19**: 161
[53] Chen C-S, Wu S, Girdaukas G, Sih CJ (1987) J Am Chem Soc **109**: 2812
[54] Chen C-S, Fujimoto Y, Girdaukas G, Sih CJ (1982) J Am Chem Soc **104**: 7294; Straathof AJJ, Jongejan JA (1997) Enzyme Microb Technol **21**: 559
[55] Anthonson T, Jongejan JA (1997) Solvent Effect in Lipase-Catalyzed Racemate Resolution. In: Rubin B, Dennis EA (eds) Methods in Enzymology, vol 286. Academic Press, New York, p 473
[56] Gotor V, Brieva R, Rebolledo F (1988) Tetrahedron Lett **29**: 6973
[57] Rakels JLI, Schneiders AGM, Straathof AJJ, Heijnen JJ (1996) Biocatal Biotransform **13**: 179
[58] Kirchner G, Scollar M, Klibanov AM (1985) J Am Chem Soc **107**: 7072
[59] Gubicza L (1992) Effect of the Solvent on Enzyme Enantioselectivity. In: Tramper J, Vermüe MH, Beeftink HH, Von Stockar U (eds) Biocatalysis in Non-Conventional Media. Elsevier Science, Amsterdam, p 497
[60] Rakels JLI, Straathof AJJ, Heijnen JJ (1994) Tetrahedron Asymm **5**: 93
[61] Starmans WAJ (1997) Synthesis and application of functionalized azetidines, Thesis, Catholic University of Nijmegen, chapter 2
[62] Starmans WAJ, Doppen RG, Thijs L, Zwanenburg B (1998) Tetrahedron Asymm **9**: 429
[63] McConville FX, Lopez JL, Wald SA (1990) Enzymatic Resolution of Ibuprofen in a Multiphase Membrane Reactor. In: Abramowicz DA (ed) Biocatalysis. Van Nostrand Reinhold, New York, p 167
[64] De Zoete MC, Ouwehand AA, Van Rantwijk F, Sheldon RA (1995) Recl Trav Chim Pays-Bas **114**: 171
[65] Hacking MAPJ, Wegman MA, Rops J, Van Rantwijk F, Sheldon RA (1998) J Mol Catal B: Enzymatic **5**: 155
[66] Conde S, López-Serrano P, Martínez A (1998) Biotechnol Lett **20**: 261
[67] Adamczyk M, Grote J, Rege S (1997) Tetrahedron Asymm **8**: 2509
[68] Sánchez VM, Rebolledo F, Gotor V (1997) Tetrahedron Asymm **8**: 37
[69] García MJ, Rebolledo F, Gotor V (1992) Tetrahedron Asymm **3**: 1519
[70] García MJ, Rebolledo F, Gotor V (1993) Tetrahedrom Asymm **4**: 2199
[71] García-Urdiales E, Rebolledo F, Gotor V (1999) Tetrahedron Asymm **10**: 721
[72] Puertas S, Rebolledo F, Gotor V (1996) J. Org Chem **61**: 6024
[73] Pozo M, Gotor V (1993) Tetrahedron **49**: 10725
[74] Kitaguchi H, Fitzpatrick PA, Huber JE, Klibanov AM (1989) J Am Chem Soc **111**: 3094; Ke T, Klibanov AM (1998) J Am Chem Soc **120**: 4259
[75] a) Kazlauskas RJ, Weissfloch ANE, Rappaport AT, Cuccia LA (1991) J Org Chem **56**: 2656; b) Cygler M, Grochulski P, Kazlauskas RJ, Schrag JD, Bouthillier F, Rubin B, Serreqi, AN, Gupta AK (1994) J Am Chem Soc **116**: 3180
[76] Kazlauskas RJ (1994) Trends Biotechnol **12**: 464 and pertinent references cited in [75b]
[77] Uppenberg J, Öhrner N, Norin M, Hult K, Kleywegt GJ, Patkar S, Waagen V, Anthonson T, Jones TA (1995) Biochemistry **34**: 16838
[78] Reetz MT, Dreisbach C (1994) Chimia **48**: 570
[79] Jaeger K-E, Liebeton K, Zonta A, Schimosschek K, Reetz MT (1996) Appl Micriobiol Biotechnol **46**: 99
[80] Pozo M, Gotor V (1993) Tehrahedron **49**: 4321
[81] Öhrner N, Orrenius C, Mattson A, Norin T, Hult K (1996) Enzyme Microbiol Technol **19**: 328
[82] Reeve CD (1999) Int Pat Appl WO 99/31264

[83] Takayama S, Lee ST, Hung S-C, Wong C-H (1999) Chem Commun (Cambridge) 127
[84] Balkenbohl F, Ditrich K, Hauer B, Ladner W (1997) J Prakt Chem **339**: 381
[85] Chulalaksananukul W, Condoret JS, Delorme P, Willemot RM (1990) FEBS **276**: 181; Rizzi M, Stylos P, Riek A, Reuss M (1992) Enzyme Microb Technol **14**: 709; Maximo MF, Van der Lugt JP (1994) Biocatalysis **8**: 321; Öhrner N, Martinelle M, Mattson A, Norin T, Hult K (1994) Biocatalysis **9**: 105; De Goede ATJW, Van Oosterom M, Van Deurzen MPJ, Sheldon RA, Van Bekkum H, Van Rantwijk F (1994) Biocatalysis **9**: 145; Berglund P, Holmquist M, Hult K, Högberg H-E (1995) Biotechnol Lett **17**: 55; Van der Heijden AM, Zuijderduijn FJ, Van Rantwijk F (1998) J Mol Catal A: Chemical **134**: 259.
[86] Reetz MT, Schimossek K (1996) Chimia **50**: 668
[87] Messina F, Botta M, Corelli F, Schneider MP, Fazio F (1999) J Org Chem **64**: 3767
[88] Nakai K, Hiratake J, Oda J (1992) Inst Chem Res Kyoto Univ **70**: 333
[89] Lopez Serrano P, Van Rantwijk F, Sheldon RA (in prep)
[90] Maestro A, Astorga C, Gotor V (1997) Tetrahedron Asymm **8**: 3153
[91] Stürmer R (1999) Eur Pat Appl EP 890649 A1
[92] Pozo M, Gotor V (1995) Tetrahedron Asymm **6**: 2797
[93] Sànchez VM, Rebolledo F, Gotor V (1999) J Org Chem **64**: 1464

Received December 20, 1999. Accepted January 1, 2000

Invited Review

Chemoenzymatic Synthesis of Lipidated Peptides

Dieter Kadereit and **Herbert Waldmann***

Max Planck Institute of Molecular Physiology, Department of Chemical Biology, D-44227 Dortmund, Germany, and University of Dortmund, Department of Organic Chemistry, D-44227 Dortmund, Germany[a]

Summary. This review highlights the use of enzymatic protecting group techniques in the synthesis of lipidated peptides. Lipidated proteins play key roles in signal transduction processes. Moreover, structurally well-defined peptides containing the characteristic linkage region of the peptide backbone with the lipid can provide valuable tools for the study of biological phenomena associated with these protein conjugates. The multifunctionality and pronounced lability towards acids and bases of such compounds render their synthesis a formidable challenge. However, the recent development of enzymatic protection groups provides an efficient access to these sensitive and biologically relevant peptide conjugates under particular mild conditions and with high selectivity.

Keywords. Lipidated peptides; Chemoenzymatic Synthesis; Enzyme-labile protecting groups; Ras proteins.

Introduction

Lipidation of proteins was discovered only two decades ago, and to date only three different types of lipid groups have been found to react with proteins by myristoylation of the N-terminal amino group [1], S-prenylation (farnesyl- and geranylgeranyl groups) of cysteine residues at or close to the C-terminus, and S-palmitoylation [1, 2] of cysteines throughout the chain (Scheme 1).

Lipid modified proteins are often attached to cell membranes. In many cases, they play crucial roles in the transduction of extracellular signals across the plasma membrane and into the nucleus. Particularly important examples are the N-, K-, and H-Ras proteins. All Ras proteins terminate in a farnesylated cysteine methyl ester. In addition, fully modified N-Ras and K-Ras$_B$ are palmitoylated at a cysteine residue close to the C-terminus, whereas H-Ras is palmitoylated twice. K-Ras$_A$ is not palmitoylated, but carries a polylysine sequence close to the farnesylated cysteine methyl ester which enhances binding to the plasma membrane (Scheme 2).

* Corresponding author

[a] Both addresses valid for both authors

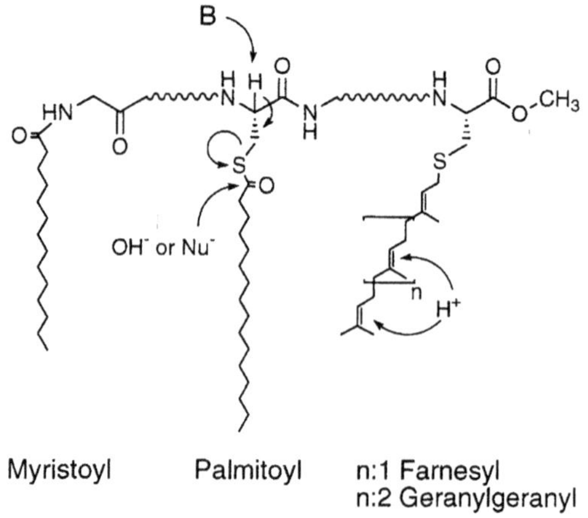

Scheme 1. Lipid modifications of proteins

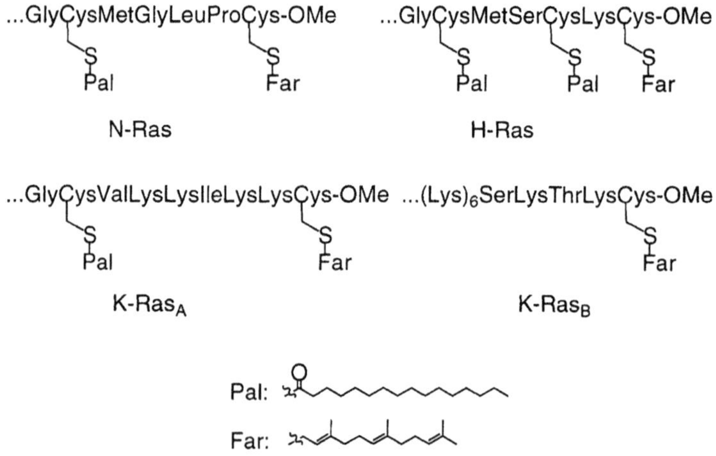

Scheme 2. Structure of lipid-modified C-termini of the Ras proteins

Lipid modification is essential for both membrane association and biological function of all Ras proteins. Non-lipidated Ras is biologically inactive.

Signal Transduction *via* Ras Proteins

Ras proteins influence numerous signal transduction processes and function as molecular binary switches which are activated by exchange of *GDP* for *GTP* and deactivated by hydrolysis of the γ-phosphate group in *GTP* to regenerate *GDP* [3].

In the Ras signal transduction cascade, monomeric receptor tyrosine kinases dimerize upon binding to an extracellular ligand, e.g. a growth factor. Then the monomer units activate each other by cross phosphorylation (Scheme 3). The phosphorylated receptors are subsequently recognized by adaptor molecules which link the receptor to signal transducers. Of particular importance is the growth factor receptor binding protein (Grb2). On the one hand, Grb2 binds *via* a SH2 domain to

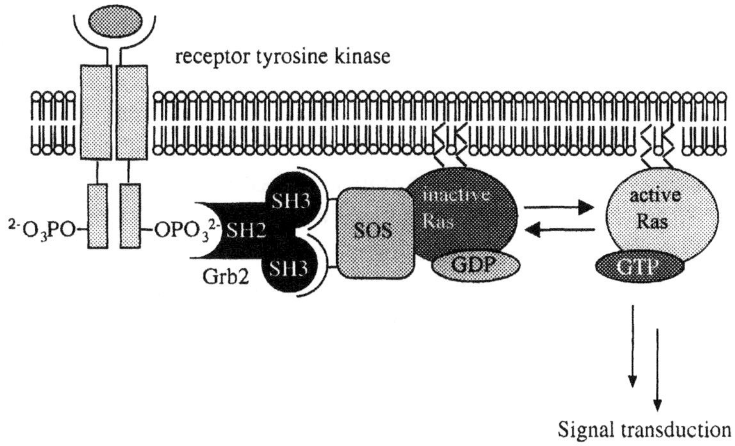

Scheme 3. Ras activation by receptor tyrosine kinases; SH2: Src (Sarcoma) Homology 2, SH3: Src (Sarcoma) Homology 3, SOS: Son of Sevenless

the receptor peptide sequence containing the phosphotyrosine moiety. On the other hand, it recognizes proline-rich sequences in another adaptor protein named SOS by means of two SH3 domains. Thereby, the cytosolic proteins Grb2 and SOS are localized and correctly aligned on the inner side of the cell membrane. SOS then interacts with the inactive *GDP*-bound form of Ras and activates it by mediating exhange of *GDP* for *GTP*. Activated *GTP*-Ras then binds to further proteins like Raf, thereby passing the signal on.

The Ras signal transduction cascade is of extreme physiological importance. It is central to the regulation of cell growth and differentiation, and false regulation of this signal pathway can affect one of the critical steps leading to cell transformation [4]. The Ras pathway is highly conserved in differen species, and its elements are used in the same way for transmission of growth signals in, for example, yeast, worms, flies, and mammals.

A mutation in Ras genes might cause a high *GDP-GTP* exchange rate or a suppressed *GTP*ase activity. Both kinds of mutations lead to continuously active Ras proteins emitting a permanent growth signal that can result in tumor formation. This malfunction is serious as statistical numbers exemplify: a mutation in Ras is found in approximately 30% of all human cancers, and in some of the major malignancies it reaches 80% [5]. The involvement of Ras genes in human cancer is not limited to their activation by point mutations. It is likely that expression of abnormally high levels of normal Ras products may also contribute to malignancy.

Although the Ras protein and its signalling activity are the subject of intense research activities, the possible functions of the lipid groups in Ras biology remain largely unclear, in particular their role in the selective targeting of Ras to the plasma membrane and the possible involvement in interactions with upstream or downstream effectors. For studying these problems, Ras peptides embodying the correct lipid parts as well as analogs thereof may be efficient tools. Further functionalities may be also required by which the peptides can be traced in biological systems. For their synthesis, enzymatic protecting group techniques have proven to be a key technology. The development of these enzyme-labile blocking functions for the construction of Ras peptides and analogs thereof is the subject of this review.

Enzyme-labile Protecting Groups in the Synthesis of Lipidated Peptides

For the synthesis of lipidated peptides and proteins, the lability of the thioester under basic conditions and towards nucleophiles poses a considerable problem. Furthermore, under acidic conditions addition of the acid to the prenyl group double bonds readily occurs. The combination of both lipid modifications, prenyl and S-acyl groups, dramatically limits the number of usable protecting groups. The acid lability of the prenyl groups (Scheme 1) excludes the application of *tert*-butyl and benzyl-type functions. Similarly, the use of *Fmoc* and related groups is ruled out due to the base lability of thioesters. For the synthesis of peptides embodying an acid-labile farnesyl thioether and a base-labile thioester, new protecting groups are required which can be removed under extremely mild, preferably neutral, conditions.

A possible solution is the use of enzyme labile protecting groups [6], since enzymatic transformations often can be carried out under advantageous and characteristically mild reaction conditions ($pH = 6$–8, room temperature to 40°C). In addition, enzymes often combine a high specificity for the recognized substrates with a large tolerance for secondary structures. Alternatively, noble metal transformations offer reaction conditions that are also mild enough to not react with sensitive, doubly lipidated peptides [7].

Enzymatic C-terminal deprotection

For the selective C-terminal deprotection of acid- and base-sensitive lipidated peptides the choline ester moiety was introduced as an enzyme-labile blocking function that can be removed under vey milk conditions [8]. Choline esters of simple peptides, but also of sensitive peptide conjugates like phosphorylated and glycosylated peptides [9], nucleopeptides [10], and also lipidated peptides [8, 11], can be cleaved with choline esterases under virtually neutral conditions. Both acetyl choline esterase (AChE) and butyryl choline esterase (BChE) can be employed for this purpose. As a rule, the butyryl choline esterase catalyzed deprotections proceed faster and result in higher yields. The high specificity of both enzymes for the choline group guarantees that only choline esters are attacked, and complete chemoselectivity is achieved. The conditions for this enzymatic deprotection are so mild that neither acid-labile farnesyl groups nor base-sensitive thioesters are attacked.

Amino acid choline esters can be synthesized readily by treatment of the corresponding 2-bromoethyl esters with trimethylamine. The charged choline esters have a pronounced solubility in aqueous solvents, a highly desirable property required for the unmasking of otherwise fairly hydrophobic lipopeptides.

For instance, in a synthesis of N-Ras lipopeptide **9**, the 2-bromoethyl ester **1** was converted into the corresponding choline ester by treatment with NMe_3. It was then removed by treatment with butyryl choline esterase (BChE) in high yield. C-Terminal elongation with S-farnesylated cysteine methyl ester using 2-ethoxy-N-ethoxycarbonyl-1,2-dihydroquinoline (*EEDQ*) as a condensing reagent followed by palladium catalyzed *Aloc* cleavage gave access to farnesylated tripeptide **5**. Similarly, 2-bromoethyl ester **6** was converted to the corresponding choline ester. Then the *Boc* group was removed, and the resulting selectively unmasked dipeptide choline ester was coupled to S-palmitoylated allyloxycarbonyl protected cysteine.

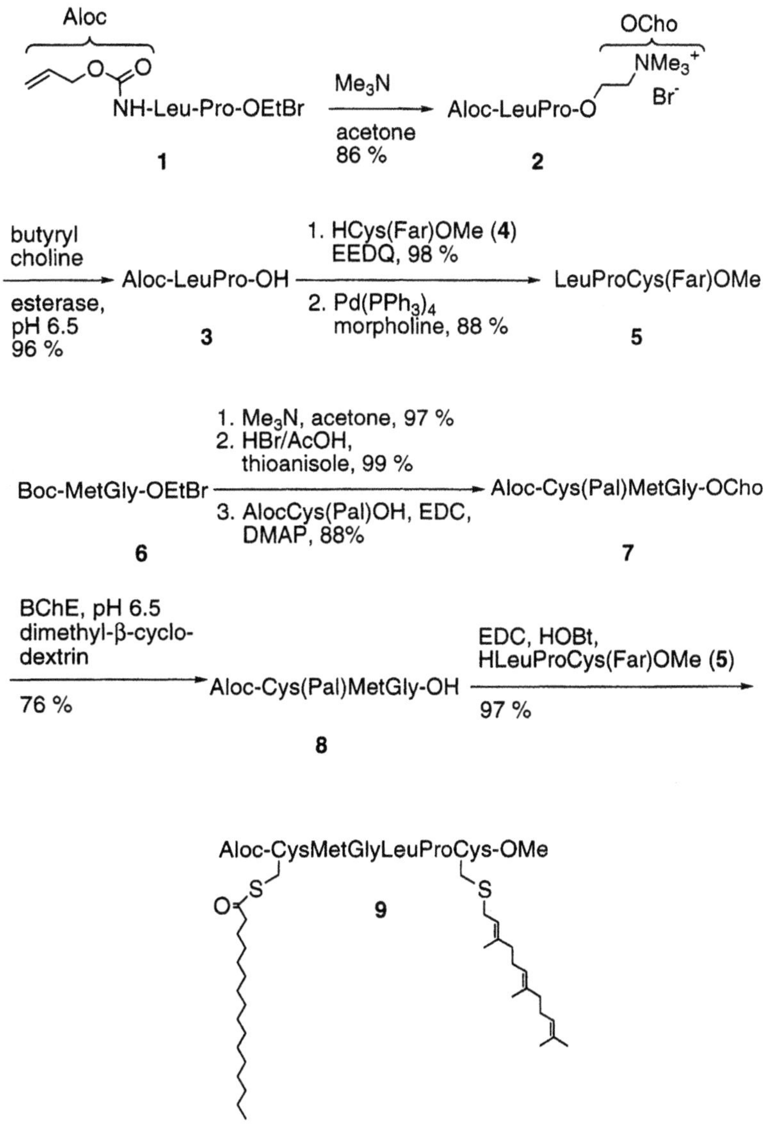

Scheme 4. N-Ras synthesis employing the choline ester as C-terminal protecting group; *EEDQ*: 2-ethoxy-N-ethoxycarbonyl-1,2-dihydroquinoline, *DMAP*: dimethylaminopyridine, *EDC*: 1-ethyl-3-(dimethylamino)-propylcarbodiimide hydrochloride, HO*Bt*: 1-hydroxybenzotriazole

From the palmitoylated base-labile tripeptide **7**, the choline ester was removed selectively and in high yield by means of the enzyme in the presence of dimethyl-β-cyclodextrin. Under these conditions, the normally chemically reactive thioester is not attacked. Thus, the biocatalyst reverses the usually observed chemoselectivity. The optimization of this deprotection step proved to be a formidable challenge. Peptide choline esters are usually highly soluble in water; thus, the substrates become readily accessible to the biocatalyst, and the use of additional solubilizing cosolvents that might denature the enzyme may be reduced or rendered unnecessary. However, the S-palmitoylated choline ester **7** is only sparingly soluble in purely aqueous media. Initial experiments using 5% of organic

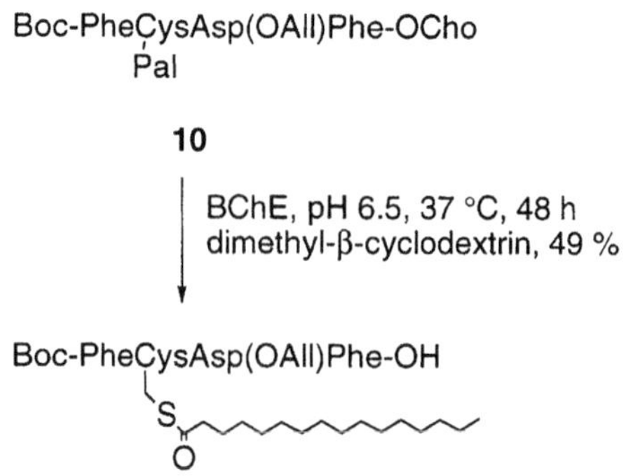

11

Scheme 5. Enzyme catalyzed deprotection of a palmitoylated Y_1 receptor peptide

cosolvents resulted in low yields, probably due to denaturation. Furthermore, in the presence of methanol, the biocatalyst catalyzes a transesterification to yield the undesired methyl ester. The addition of dimethyl-β-cyclodextrin instead of standard organic cosolvents, however, resulted in an enhanced solubility of peptide **7** and in a smooth conversion to the free acid **8**. Cyclodextrins are cyclic heptasaccharides with a hydrophobic cavity. It is assumed that the cyclodextrines can slip over the hydrophobic lipid residues, thereby shielding them from the solvent. In addition, this shielding may also prevent non-enzymatic hydrolysis of the thioesters. Efficient coupling of both lipidmodified tripeptides **5** and **8** in high yield completed the synthesis of the target peptide **9**.

Similarly, cyclodextrins were successfully applied in a synthesis of palmitoylated tetrapeptide **11** [12]. Tetrapeptide choline ester **10** was deprotected by BChE in the presence of dimethyl-β-cyclodextrin. In the absence of the solubility enhancer this reaction did not proceed.

A similar strategy was applied in the synthesis of the hexapeptide **16** which represents the characteristic N-myristoylated and S-palmitoylated amino terminus of the $G_{\alpha 0}$-protein [11].

The synthesis of the doubly lipidated peptide **16** proceeded *via* enzyme catalyzed deprotection of the palmitoylated tetrapeptide choline ester **12** in the presence of cyclodextrin. C-Terminal elongation with dipeptide choline ester **14** gave the corresponding hexapeptide **15**. Once again treatment of **15** with BChE in the presence of the solubility enhancing cyclic heptasaccharide resulted in smooth hydrolysis of the choline ester without affecting the thioester. Finally, treatment with trifluoroacetic acid (*TFA*) followed by introduction of the myristoyl group gave the fully modified hexapeptide **16** in good yield.

However, in severe cases, even the combined use of choline esters and cyclodextrins can not overcome inherent solubility problems. This observation was particularly present in various N-myristoylated lipopeptides. For instance, initially the synthesis of **16** was attempted by starting with a previously myristoylated analogue of **12**. The doubly lipidated peptides were only slightly soluble in the aqueous buffer even in the presence of cyclodextrins, and no enzymatic deprotection could be observed. Addition of organic cosolvents did not improve the deprotection but rather resulted in denaturation of the enzyme.

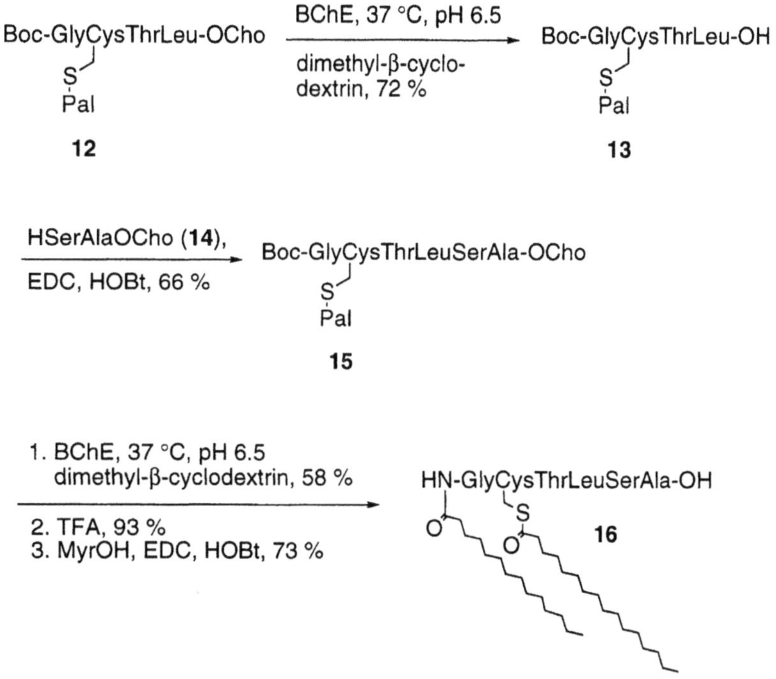

Scheme 6. Choline ester strategy for the synthesis of a myristoylated and palmitoylated hexapeptide corresponding to the $G_{\alpha 0}$-protein N-terminus; *TFA*: trifluoroacetic acid, *EDC*: 1-ethyl-3-(dimethylamino)-propylcarbodiimide hydrochloride, HO*Bt*: 1-hydroxybenzotriazole

Enzyme labile amine protecting groups

The development of enzyme-labile protecting groups for the N-terminus of peptides and peptide conjugates poses a major challenge. First, these blocking functions have to be removable under mild conditions to prevent base mediated thioester hydrolysis as well as acid catalyzed addition to the prenyl double bonds. In addition, they should be removable with a biocatalyst which does not attack other functional groups present in the molecule, especially the palmitoyl thioesters. Finally, they have to embody a function group which is specifically recognized by the enzyme, preferably an urethane structure to avoid racemization upon amino acid activation. Unfortunately, most enzymes available today do not attack urethanes which might be due to diminished reactivity of the urethane carbonyl group. An alternative strategy would be to employ a biocatalyst which attacks a different bond, *e.g.* an O-alkyl or an ester bond, and to design such a urethane accordingly.

For instance, the design of the carbohydrate based tetrabenzylglucosyloxy-carbonyl (*BGloc*) protecting group is based on the ability of glycosidases to hydrolyze the glycosidic bond of the corresponding sugar, the bond between C-1 and O-1 [13]. In this case, the enzyme (here: a glucosidase) initially cleaves the glycosidic C-O bond to the carbohydrate thereby rendering an attack on the less reactive urethane unnecessary (Scheme 7).

The removal of *BGloc* proceeds *via* a two step chemoenzymatic protocol. First, in the chemical step, the benzyl ethers are cleaved by hydrogenolysis. Second, enzaymatic treatment with a glucosidase to yield glucose, CO_2, and the desired free

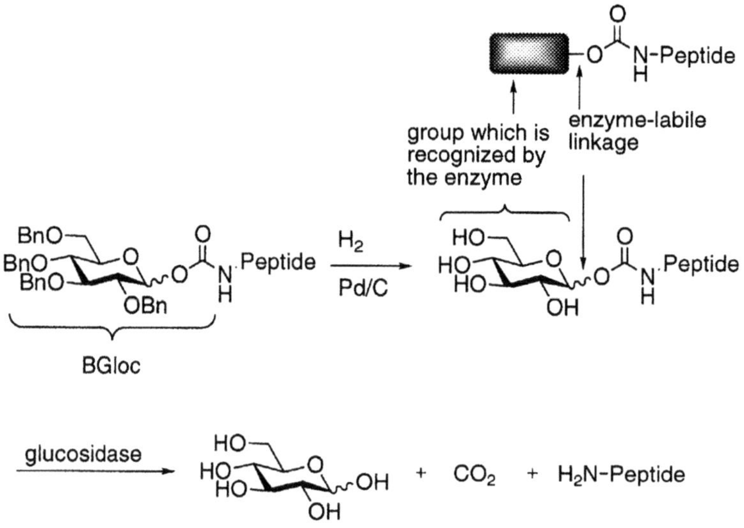

Scheme 7. The *BGloc* protecting group

amine was achieved. However, such carbohydrate based protecting group have not yet been used in the synthesis of lipidated peptides.

Another general strategy for the design of enzymatically removable protecting groups is to link the urethane *via* a spacer to the functional group which is

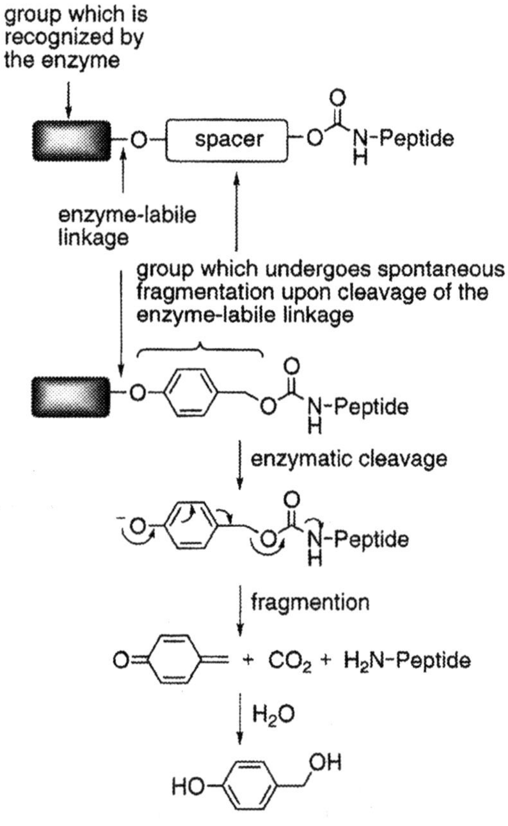

Scheme 8. Enzymatically removable protecting groups embodying a fragmenting spacer unit

specifically recognized by the enzyme. Upon cleavage of the enzyme sensitive bond, the spacer undergoes spontaneous fragmentation. In this fragmentation process, a carbamic acid derivative is liberated which decarboxylates to finally yield the desired peptide or peptide conjugate (Scheme 8).

The principle of the enzymatic deprotection depicted in Scheme 8 is general. Depending on the acyl group chosen, the fragmentation of the resulting *p*-acyloxybenzyl urethane can be initiated with an appropriate enzyme. An additional advantageous feature is that the variable peptide part of the substrate is remote from the site of the biocatalysis. Thus, possible unfavourably steric or electronic interactions of the protein with the peptide caused by bulky amino acid side are minimized. Therefore, this enzymatic protecting group technique can be applied for the construction of peptides and analogues thereof containing, for instance, unnatural amino acids including the *D* amino acids.

The p-acetoxybenzyloxycarbonyl group (AcOZ)

The *p*-acetoxybenzyloxycarbonyl (AcOZ) group, originally introduced as base-labile blocking function [14], can be cleaved readily by means of lipase- or esterase- initiated fragmentation under exceptionally mild conditions (*pH* 5–6) [15, 16].

AcOZ urethanes can be removed efficiently by a lipase from *Mucor miehei* and an acetyl esterase from the *flavedo* of oranges [17]. The lipase deprotected even sterically demanding amino acids. In addition, it tolerated high amounts of methanol as a cosolvent. On the other hand, the acetyl esterase discriminated between acetyl and longer acyl side chains. This feature was useful especially for the removal of AcOZ in the presence of palmitoyl thioesters (Scheme 9).

The synthesis of the C-terminal N-Ras heptapeptide **23** proceeded *via* lipase catalyzed removal of the AcOZ group from tripeptide **17** in the presence of 20% methanol as solubilizing cosolvent [16]. During the course of this reaction, one equivalent of quinone methide was formed by fragmentation of the linker. To trap this reactive intermediate, an excess of potassium iodide was added. It should be noted that the cysteine methyl ester was not affected under these conditions. N-terminal elongation with dipeptide **19** yielded a pentapeptide that was again deprotected under lipase catalysis. Further elongation resulted in the palmitoylated and farnesylated heptapeptide **22**. Acetyl esterase catalyzed removal of the AcOZ urethane from **22** was accomplished at *pH* 6 in the presence of the base labile thioester. Dimethyl-β-cyclodextrins were added to improve solubility of peptide **22**. Although conversion of **22** was complete, due to its amphiphilic nature part of the desired heptapeptide **23** was lost during the isolation procedure.

Synthesis of Lipidated Peptides for Biological Investigations

The development of the methodologies discussed thus far has allowed for the synthesis of a variety of lipid modified peptides representing characteristic partial structures of naturally occuring lipidated proteins. However, for the study of biological phenomena, additional analogues with modified lipid or peptide structure may be required. In addition, the introduction of reporter groups, which allow for monitoring the intracellular fate of the peptide conjugates, may be necessary. Depending on the nature of the lipid group, several problems have been studied.

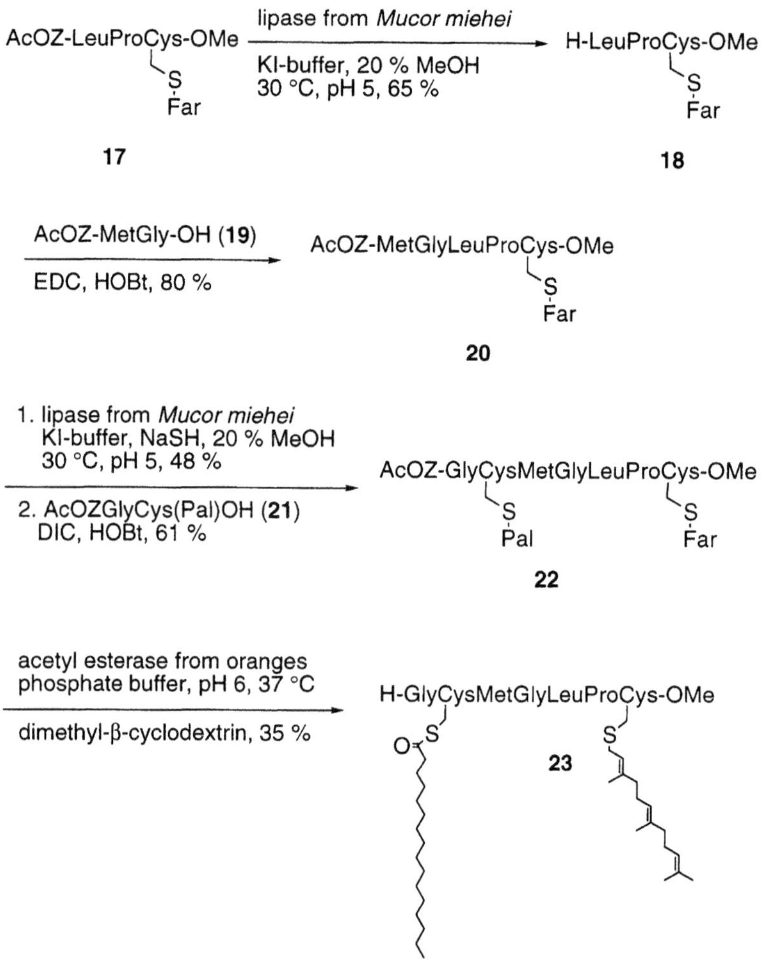

Scheme 9. Use of the AcOZ urethane in the synthesis of palmitoylated and farnesylated N-Ras heptapeptide **23**

Under physiological conditions, S-palmitoylation is a reversible process. Thus, the regulation of palmitoylation/depalmitoylation processes may be involved in the steering of biological phenomena like regulated membrane trapping mechanisms. To investigate such mechanisms, peptides are needed which either can not be palmitoylated (*i.e. Cys→Ser* or *Cys→Ala* mutants) or which are irreversibly modified, i.e. embodying a cysteine hexadecyl thioether instead of the corresponding palmitoyl thioester. In the case of the N-Ras C-terminus and the N-terminal sequence of the human $G_{\alpha 0}$-protein (Scheme 10), such analogues were synthesized.

Hexadecyl thioethers were synthesized by alkylation of mercapto groups with hexadecyl bromide [8b,11,16,18]. If free thiol groups in peptides are generated for or during a biological experiment, a suitable protecting group (e.g. *tert*-butyl disulfide) must be introduced which can be removed during the final steps of the synthesis.

Similarly, the farnesyl group in proteins may participate in protein-protein interactions. Even more so, the existence of farnesyl receptors in membranes is believed to be possible. On the other hand, only the hydrophobicity of that lipid group might account for its physiological consequences. Thus, for the investigation

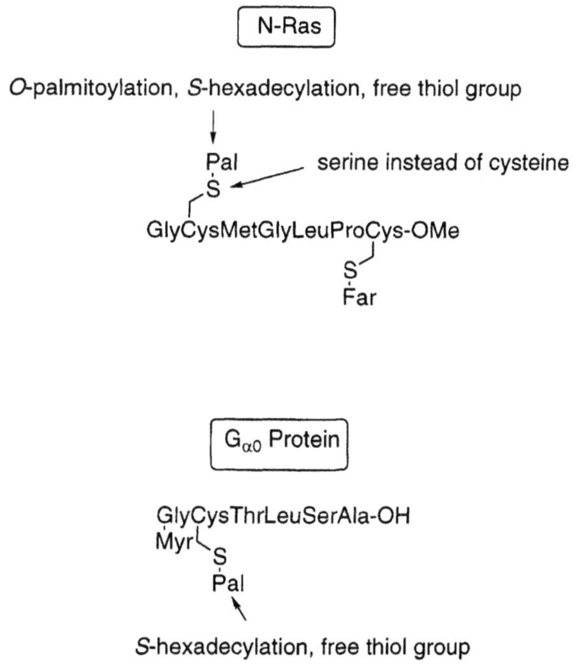

Scheme 10. Variation of S-lipidation and amino acid residues for the investigation of the biological function of protein palmitoylation

of the biological importance of protein farnesylation, analogues are needed which display a similar hydrophobicity but different structure (*e.g. n*-alkyl ethers). Other analogues, especially in the case of palmitoylated Ras proteins, contain either an unmodified cysteine or a cysteine which has been replaced for a serine [19].

For biological assays, lipidated peptides embodying a fluorescent label like the bimanyl-, the *NBD*-group, or fluoresceine are required for determining cell uptake, membrane binding or subcellular distribution by fluorescence spectroscopy and fluorescence microscopy. Also, attachment of a biotin group allows researches to trace such modified peptides by means of the protein streptavidine, which may carry a fluorescent lable or gold clusters [30]. In general, such functional groups can be attached to the amino group of selectively N-terminal deprotected

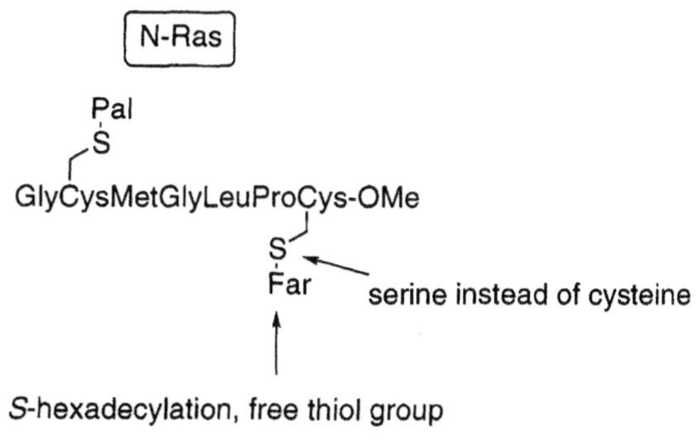

Scheme 11. Peptides for the investigation of the biological function of the farnesyl thioether

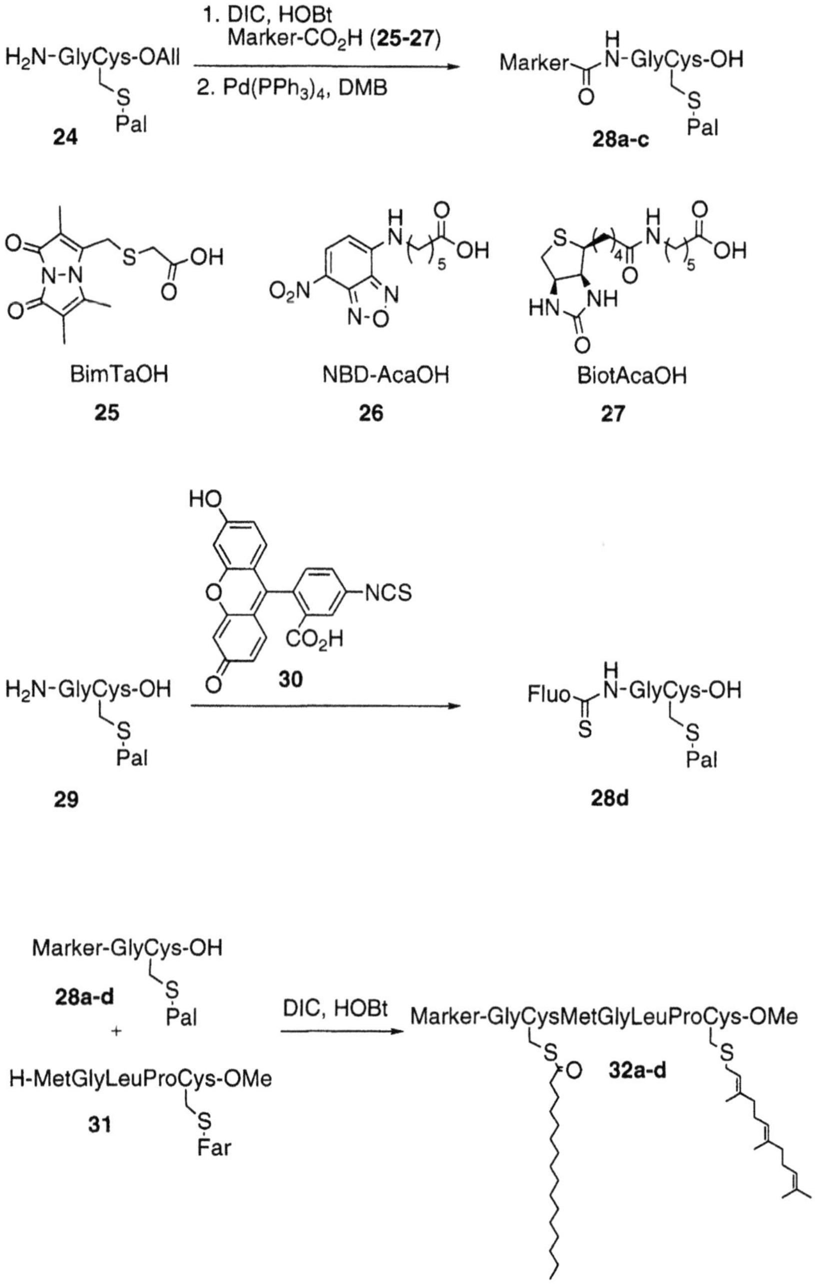

Scheme 12. Synthesis of lipidated peptides carrying a fluorescent label at the N-terminus; *DIC*: diisopropylcarbodiimide, HO*Bt*: N-hydroxybenzotriazole, *DMB*: dimethylbarbituric acid

peptides like **24** (Scheme 12). Thus, lipidated peptide **24** was treated with S-bimanylthioacetic acid (*BimTa*OH, **25**), 7-nitrobenzofurazene-4-aminocaproic acid (*NBD-Aca*OH, **26**), or fluoresceine isothiocyanate (**30**) to afford fluorescent-labelled peptides **28** [16]. Accordingly, coupling with biotinylaminocaproic acid (**27**) yielded biotinylated peptides like **28c** [8b]. Condensation of these dipeptides with farnesylated peptide **31**, which is available *via* AcOZ-technology (Scheme 9), led to the formation of labelled C-terminal N-Ras peptides.

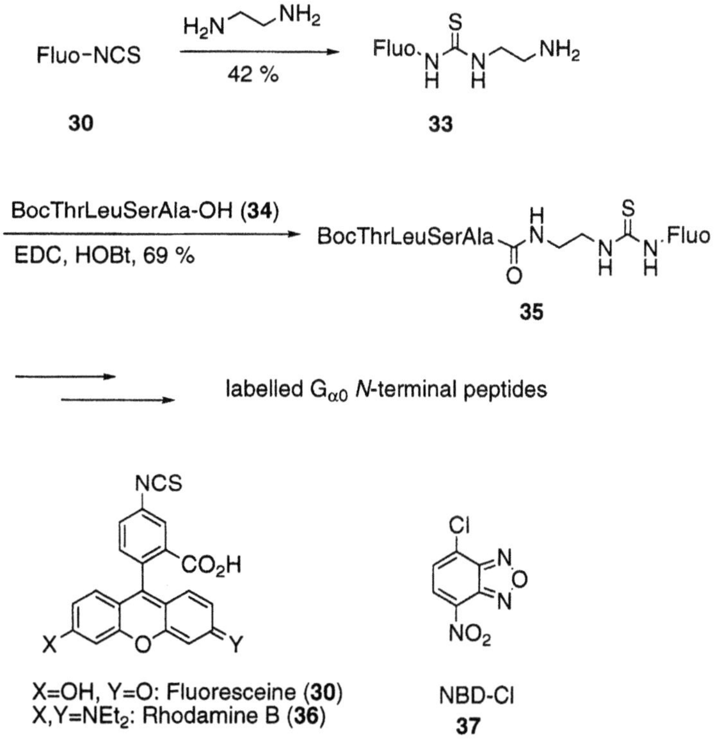

Scheme 13. Synthesis of lipidated peptides with a fluorescent marker at the C-terminus

For the introduction of fluorescent markers at the C-terminal carboxy group in peptides, fluoresceine isothiocyanate (**30**) was treated with ethylenediamine (Scheme 11). The amino-functionalized label was then condensed with peptide **34** followed by an N-terminal deprotection/chain elongation process [11]. Accordingly, rhodamine B isothiocyanate (**36**) and 4-chloro-7-nitrobenzofurazene (*NBD*-Cl, **37**) were converted into the respective ethylenediamine derivative and were attached to the peptide carboxyl groups.

Lipidated peptides embodying the characteristic linkage region found in the parent lipoproteins and bearing additional functional groups which could be traced in biological systems or which allowed for their use in biophysical experiments were used successfully in model studies. The application of lipid modified conjugates in binding experiments to vesicles and model membranes, in membrane fusion experiments, and in microinjection studies has led to the proposal of a mechanism for the targeting of Ras proteins to the plasma membrane [17, 21].

For discussion of these experiments and the consequences for targeting of Ras and related proteins to the plasma membrane the reader is referred to more comprehensive review [22].

Conclusions and outlook

The enzyme-mediated syntheses of lipidated peptides described in the previous sections clearly demonstrate the capabilities of enzymatic transformations, in particular enzymatic protecting group techniques. Today, an increasing number of biological phenomena is being investigated, elucidated, and understood in molecular detail at a rapidly increasing speed. For organic chemistry, this opens up a multitude of new spheres of activity in which its power and strength can be used to the full extent and in which new, great, and important challenges are presented. In mastering them, organic chemistry can rise to a key role.

Acknowledgements

We would like to thank the *Deutsche Forschungsgemeinschaft* and the *Fonds der Chemischen Industrie* for generous support of our work.

References

[1] Resh MD (1999) Biochim Biophys Acta **1451**: 1
[2] Dunphy JT, Linder ME (1998) Biochim Biophys Acta **1436**: 245
[3] For recent reviews see (a) Geyer M, Wittinghofer A (1997) Curr Op Struct Biol **7**: 786; (b) Rommel C, Hafen E (1998) Curr Op Gen Dev **8**: 412; (c) Vojtek AB, Der CJ (1998) J Biol Chem **273**: 19925
[4] Waldmann H, Wittinghofer A, Angew Chem (in press)
[5] Barbacid M (1987) Annu Rev Biochem **56**: 779
[6] (a) Waldmann H, Sebastian D (1994) Chem Rev **94**: 911; (b) Schelhaas M, Waldmann H (1996) Angew Chem **108**: 2192, Angew Chem Int Ed Engl **35**: 2056; (c) Pathak T, Waldmann H (1998) Curr Op Chem Biol **2**: 112
[7] (a) Guibé F (1997) Tetrahedron **53**: 13509; (b) Guibé F (1998) Tetrahedron **54**: 2967
[8] (a) Schelhaas M, Glomsda S, Hänsler M, Jakubke HD, Waldmann H (1996) Angew Chem **108**: 82, Angew Chem Int Ed Engl **35**: 106; (b) Schelhaas M, Nägele E, Kuder N, Bader B, Kuhlmann J, Wittinghofer A, Waldmann H (1999) Chem Eur J **5**: 1239
[9] Sander J, Waldmann H (1999) Angew Chem **111**: 1337, Angew Chem Int Ed Engl **38**: 1250
[10] Flohr S, Jungmann V, Waldmann H (1999) Chem Eur J **5**: 669
[11] Cotté A, Bader B, Kuhlmann J, Waldmann H (1999) Chem Eur J **5**: 922
[12] Kuhn K, Waldmann H (1999) Tetrahedron Lett **40**: 6369
[13] Kappes T, Waldmann H (1998) Carbohydr Res **305**: 341
[14] Le Corre G, Guibé-Jampel E, Wakselman M (1978) Tetrahedron **34**: 3105
[15] Waldmann H, Nägele E (1995) Angew Chem **107**: 2425, Angew Chem Int Ed Engl **34**: 2259
[16] Nägele E, Schelhaas M, Kuder N, Waldmann H (1998) J Am Chem Soc **120**: 6889
[17] Waldmann H, Schelhaas M, Nägele E, Kuhlmann J, Witinghofer A, Schroeder H, Silvius JR (1997) Angew Chem **109**: 2334, Angew Chem Int Ed Engl **36**: 2238
[18] Koppitz M, Spellig T, Kahmann R, Kessler H (1996) Int J Peptide Protein Res **48**: 377
[19] Owen DJ, Kuhn K, Bader B, Wittinghofer A, Kuhlmann J, Waldmann H, Angew Chem (submitted for publication)
[20] Hermanson GT (1996) In: Bioconjugate Techniques. Academic Press, London
[21] Schroeder H, Leventis R, Rex S, Schelhaas M, Nägele E, Waldmann H, Silvius JR (1997) Biochemistry **36**: 13102
[22] (a) Hinterding K, Alonso-Díaz D, Waldmann H (1998) Angew Chem **110**: 716, Angew Chem Int Ed Engl **37**: 688; (b) Kadereit D, Kuhlmann J, Waldmann H (2000) Angew Chem (submitted for publication)

Received December 12, 1999. Accepted January 26, 2000

Invited Review

Chemoenzymatic Transformations in Nucleoside Chemistry

Miguel Ferrero and **Vicente Gotor***

Departamento de Química Orgánica e Inorgánica, Facultad de Química, Universidad de Oviedo, E-33006 Oviedo, Spain

Summary. The synthetic potential of enzymes in connection with nucleoside analogue preparations is described.

Keywords. Enzymes; Nucleosides; Chemoenzymatic transformations.

Introduction

During the last quarter of the century almost no other aspect of organic synthesis has received as much attention as the preparation of enantiomerically pure compounds. The synthesis of optically active materials is an important task and represents a challenge to synthetic organic chemists. Increasing interest in the understanding of biological processes and the general recognition that chirality plays a crucial role in nature fostered a tremendous effort in enantioselective synthesis. Nowadays, the choice of a racemic synthesis over the development of an enantiopure compound must be justified, since opposite enantiomers interact differently within an organism and can display various activities. Thus, each country requires investigation into the bio-availability and pharmacological effect of a new chiral drug, and its final approval is based on complete background information for each enantiomer.

The chemistry of natural nucleosides and their analogues has been thoroughly studied due to their potential as fungicidal, antitumour, and antiviral agents [1]. Consequently, extensive modifications have been made to both the heterocyclic base and the sugar moiety in order to avoid the drawbacks shown by nucleosides or analogues in certain applications. Since the late eighties, nucleoside analogues have been investigated with renewed urgency in the search for agents effective against the Human Immunodeficiency Virus (HIV), the causative agent of AIDS. More effective treatment has also been sought for other viral infections, in particular those caused by Herpes Simplex Virus (HSV types 1 and 2), Varicella Zoster Virus (VZV), Human Cytomegalovirus (HCMV), and *Epstein-Barr* Virus

* Corresponding author

(EBV), which can prove lethal to AIDS patients and other immunocompromised individuals. This has resulted in an explosion of synthetic activity in the field of nucleosides, with special interest in their asymmetric syntheses, and in the discovery of a number of derivatives with potential antiviral activity.

For organic chemists, enzyme-catalyzed reactions have become standard procedures for the synthesis of enantiomerically pure compounds due to their simple feasibility and high efficiency. In general, these catalysts are inexpensive and in many cases able to fit to a wide range of substrate structures. Moreover, biocatalysts are ecologically beneficial natural catalysts. Due to these advantages, it is to be expected that biocatalyzed reactions will play an increasing role primarily in the preparation of nonracemic chiral biologically active compounds in the laboratory as well as in industrial production.

A number of extensive general reviews [2], special issues of periodicals [3], or specialized books [4] have been published during the last two decades on the subject of biocatalysis in organic synthesis. This is due to the explosive development and success of this methodology in the area of the synthesis of

Table 1. Enzymes commonly used in biocatalytic processes

Accepted name	Abbreviation	Other denominations
Candida rugosa lipase	CRL	*Candida cylindracea* lipase (CCL)
Candida antarctica lipase	CAL	SP-435, lipase B, CAL-B
Pseudomanas capacea lipase	PSL	*Pseudomonas sp.* lipase, *Pseudomonas fluorescens* lipase (PFL), PCL, lipase P, lipase PS, Amano PS, KWI-56
Lipase M		
Porcine pancreas lipase	PPL	
Pig liver esterase	PLE	
Acetyl esterase	AE	
α-Chymotrypsin	α-CT	
Subtilisin		
Penicillin G acylase	PGA	
Adenosine deaminase	ADA	
Adenilic acid deaminase	AMPDA	
Penicillin amidase	PA	
Cytidine deaminase	CDA	
Papain		
Wheat germ lipase	WGL	
Aspergillus niger lipase	ANL	
Butyrylcholine esterase	BCE	
β-Galactosidase		
Cyclase (*Aplysia californica*)		
Nucleoside oxidase	FERM BP-2252	
FDP aldolase		
Phosphatase		
Phosphorylase		
CloneZyme ESL-001-02	ELS-001-02	

natural products, pharmaceuticals, and agrochemicals. These reviews and monographs do not usually focus their interest on the target molecule, but rather on the structure of the substrate. Moreover, although there are several review articles dealing with natural products such as carbohydrates [5] and steroids [6], little has been reported on biotransformations in the field of nucleosides, especially from the point of view of enzymatic reactions, except for a review article about enzyme protecting groups restricted to nucleosides containing the natural pentanofuranose moiety [7] and two recent reviews, one dealing with the enzymatic preparation of nucleoside antibiotics using microorganisms [8]; and the other with nucleoside synthesis mediated by glycosyl transferring enzymes [9].

The synthetic potential of enzymes related to nucleoside synthesis has been applied increasingly, especially since the introduction of organic solvent methodology. It is the aim of this report to cover the literature from the beginning of the 80s up to the end of 1999 with respect to chemoenzymatic transformations in conventional nucleosides. It is the intention of the authors to show a range of examples that cover nucleoside analogue syntheses through enzymatic procedures. Due to the vastness of the bibliographic material related to nucleosides, we will focus our attention especially on those processes which involve lipases, esterases, and proteases. Thus, the review will not cover carbocyclic nucleosides or C-nucleosides.

Most enzymatic reactions, just like those shown in this review, are mediated by a small number of biocatalysts. With the passing of time, their nomenclature has changed in an effort to unify criteria and refer to a given enzyme by only one name. Table 1 lists the enzymes mentioned in this review. These are cited as in the original papers to facilitate checking the original work, together with their corresponding modern denominations.

Enzymatic Hydrolysis on the Base Moiety

The first enzymatic hydrolysis of 6-subtituted purine riboses (**1**) was performed by *Wolfenden* [10] in 1966 in which adenosine deaminase (ADA) catalyzed the hydrolysis of 6-hydrazinopurine ribose, 6-chloropurine ribose, 6-methylaminopurine ribose, and 6-methoxypurine ribose at limiting velocities very similar to that for adenosine itself (Scheme 1). Each of these compounds, when incubated with the purified enzyme at *pH* 6.5, was completely converted to inosine (**2**). A common rate of hydrolysis for such a variety of leaving groups strongly suggests that the

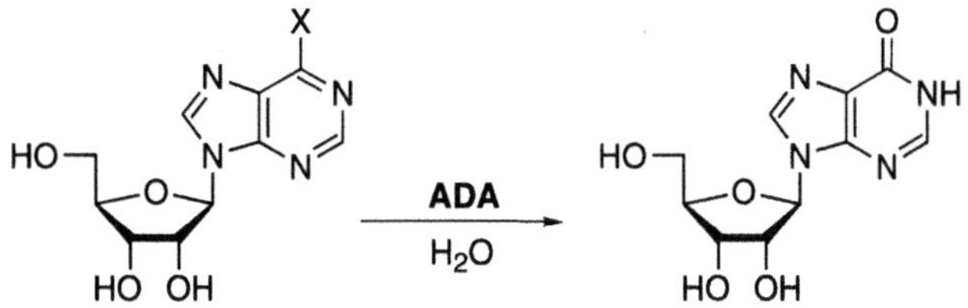

1 **2**

X = NH$_2$, NHNH$_2$, Cl, NHMe, OMe

Scheme 1

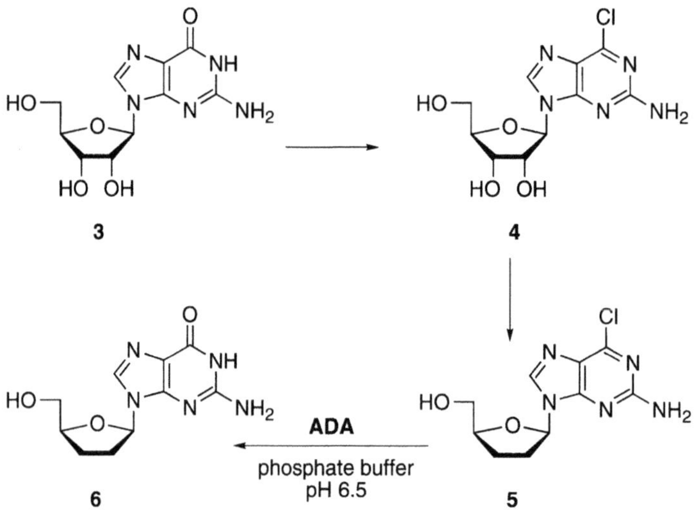

Scheme 2

same intermediate from which the original 6-substituent has been completely displaced is present in all cases. None of these reactions is detectably reversible at *pH* 6.5 or 9.

Due to its ability to inhibit the cytopathic effect of HIV, 2′,3′-dideoxyguanosine (**6**), synthesized from the readily available guanosin (**3**), is an important target molecule. Its synthesis [11] has been performed using commercially available mammalian ADA in the key step (Scheme 2). 2′,3′-Dideoxynucleoside **5** was obtained in 40% overall yield from **4** following the cyclic thionocarbonate methodology. An additional advantage of this approach is that it can be readily scaled up to gram quantities.

Despite the relatively weak reactivity of the exocyclic amino group of 2′-deoxyadenosine or 2′-deoxyguanosine under the conditions for nucleotide bond formation, its selective protection is an unavoidable step in oligonucleotide synthesis. Thus, *Galunsky et al.* [12] have reported that the phenylacetyl group can be successfully used as an enzyme cleavable aminoprotecting group of 2′-deoxyadenosine (**9**) and 2′-deoxyguanosine (**10**). These derivatives, **7** and **8**, were subjected to deprotection, catalyzed by free or immobilized penicillin amidase (PA), yielding adenosine and guanosine derivatives in 80% and 75% yield, respectively (Scheme 3). The data suggest that PA would be effective in multiple deprotection of oligonucleotides containing N-phenylacetylated purine nucleobases under mild conditions. Moreover, the acceptable deprotection times (15–150 minutes) support this conclusion.

Adenosine deaminase (ADA) catalyzes the conversion of adenosine into inosine and is physiologically important in purine metabolism. Taking advantage of this natural process, ADA has been applied in the syntheses of guanosine and inosine nucleosides from the corresponding 2,6-diamino (**12**) or 6-aminopurine nucleosides (**11**), respectively [13] (Scheme 4). ADA specifically catalyses the oxidation of a particular amino group present in the base moiety of the nucleoside, leaving the amino group of a 2′-amino sugar moiety intact as *e.g.* in compound **14** [14].

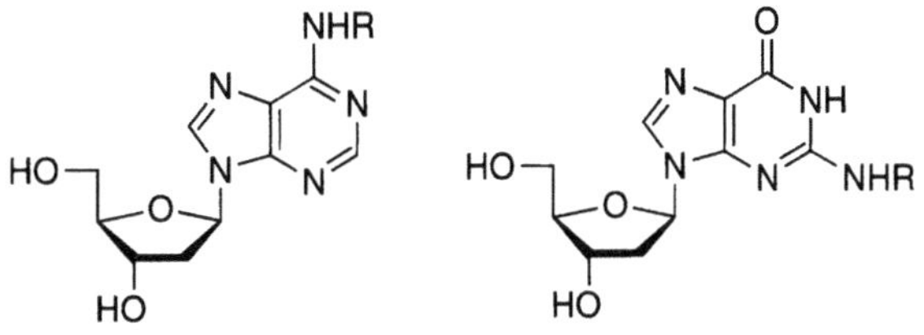

7: *R* = $COCH_2Ph$ **8:** *R* = $COCH_2Ph$

9: *R* = H (80%) **10:** *R* = H (75%)

Scheme 3

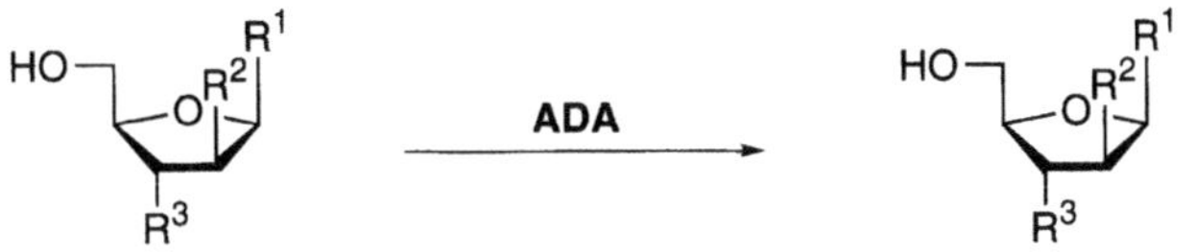

11: R^1 = H, R^2 = *B*, R^3 = OH or CH_2OH

12: R^1 = *B*, R^2 = NH_2, R^3 = OH

B = 2-aminoadenine

13: R^1 = H, R^2 = B^1, R^3 = OH or CH_2OH

14: R^1 = B^1, R^2 = NH_2, R^3 = OH

B^1 = adenine

Scheme 4

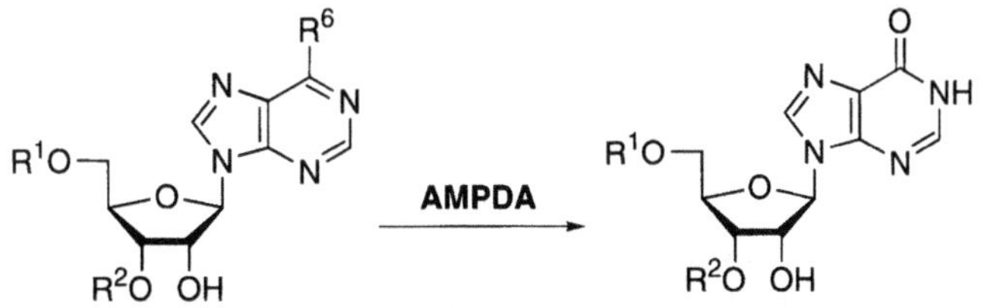

15 **16**

R^1 = H, HPO_3, R^2 = H, R^1R^2 = HPO_2, R^6 = NH_2, OMe, Cl

Scheme 5

Margolin et al. [15] have recently used adenylic acid deaminase (AMPDA) from *Aspergillus niger* in the synthesis of 6-oxopurine nucleoside in quantitative yields. This enzyme shows much broader substrate specificity and has been used for the deamination of several derivatives of adenosine **15**, including phosphorylated and cyclic analogues on a preparative scale, as well as for the dechlorination and demethylation of the purine ribosides to give the corresponding optically active 6-oxopurine derivatives **16** (Scheme 5).

In nucleotide chemistry the protection and deprotection of the amino has to be carried out under mild conditions and with a high degree of selectivity. An enzymatically removable protecting group such as phenylacetamide meets these requirements. Thus, *Waldmann et al.* [16] have shown that the phenylacetamides of 3′,5′-di-O-acetylated 2′-deoxyadenosine **17** and the respective guanosine derivative

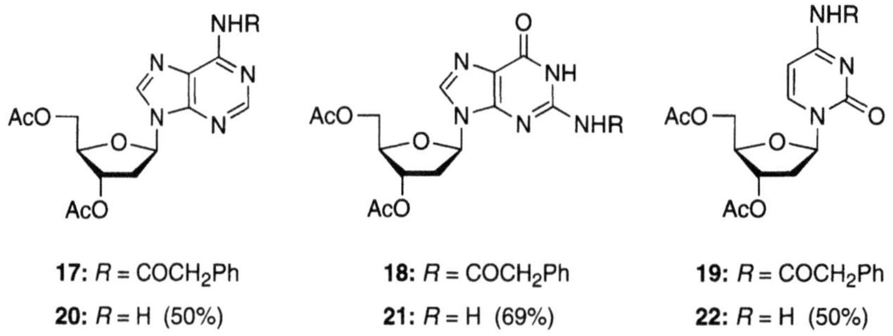

Scheme 6

18 can be prepared in a straightforward way in high yields by acylation of the corresponding amines with phenylacetic anhydride in pyridine. If they are treated with penicillin G acylase (PGA) at *pH* 7.8 in phosphate buffer, the phenylacetamido group is cleaved to deliver the selectively deprotected purine nucleobases **20** and **21** (Scheme 6). In addition, the respective protected cytidine derivative **19** is converted to the selectively deblocked pyrimidine nucleoside **22**. The conversion of the substrates is quantitative, but due to losses during work-up, the products are isolated in the yields shown in Scheme 6. The enzymatic attack on the phenylacetamides occurs with complete chemoselectitivity, leaving the more reactive esters in the carbohydrate part intact. Moreover, the reaction conditions are so mild that no undesired side reaction occurs. This new protecting group technique represents the first example for an enzymatic unmasking of amino functions in the nucleoside field. The same research team [17] has also applied this methodology to remove the phenylacetyl protecting group from oligonucleotides both in solution and on solid support.

The search for novel nucleoside structures for use as antiviral agents has lead *Van Draanen* and *Koszalka* [18] to centre their interest in α-*L*-nucleosides because of their unnatural configuration at C-4′. Their work shows the preparation and evaluation of a series of novel purine and pyrimidine α-*L*-2′,3′-dideoxynucleosides. The synthesis of pyrimidine α-*L*-2′,3′-dideoxynucleosides was effected in a straightforward manner from *D*-glutamic acid (**23**) giving rise to the target nucleosides as α,β-anomeric mixtures (**24**). The α,β-anomeric mixtures **24** were separated either by preparative HPLC or, when the uracil and thymidine analogues were used, by enzymatic resolution as shown in Scheme 7. Thus, cytidine deaminase (CDA) deaminated the α-anomer **25** much more rapidly than the β-anomer **26**. By taking advantage of this enzymatic selectivity, pure α-*L*-2′,3′-dideoxyuridine and α-*L*-3′-dideoxythymidine were prepared. These compounds were easily separated from the unreacted β-cytidine analogues by flash chromatography. The purine dideoxynucleosides **27** were synthesized by a phosphorylase-catalyzed enzymatic *trans*-ribosylation from α-*L*-dideoxyuridine. Finally, the hypoxanthine and guanine analogues **28** were prepared by treatment of the corresponding adenine and diaminopurine analogues **27** with adenosine deaminase (ADA).

As a part of continuing interest in fluorodeoxy nucleosides and nucleotides, *Mikhailopulo et al.* [19] have reported on the enzymatic synthesis of 2′-deoxy-2′-fluoroguanosine (**32**) and its antiviral activity in chick embryo cells infected with

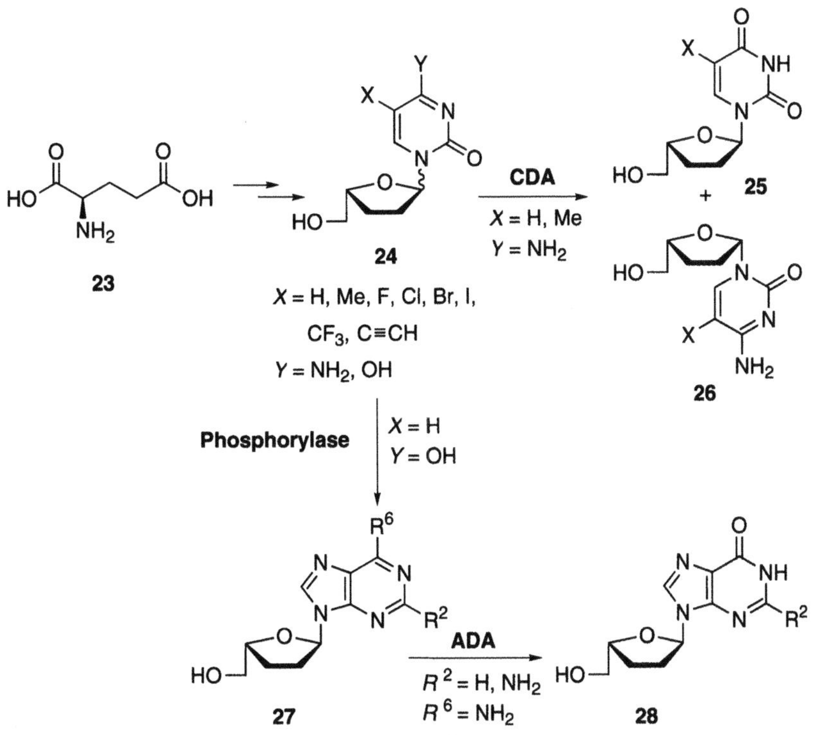

Scheme 7

Influenza Virus FPV/Rostock/34 (H7N1) and Herpes Simplex Virus (HSV) type 1 (1C strain). An enzymatic transglycosylation of 2,6-diaminopurine (**30**) using the glutaraldehyde-treated whole cells of *E. coli* BMT-4D/1A as a biocatalyst and 2′-deoxy-2′-fluorouridine or -cytidine **29** as a donor of the glycosyl moiety followed by an enzymatic deamination of the intermediate 2,6-diaminopurine glycoside **31** afforded the fluoro compound **32** in 72–79% combined yield (Scheme 8).

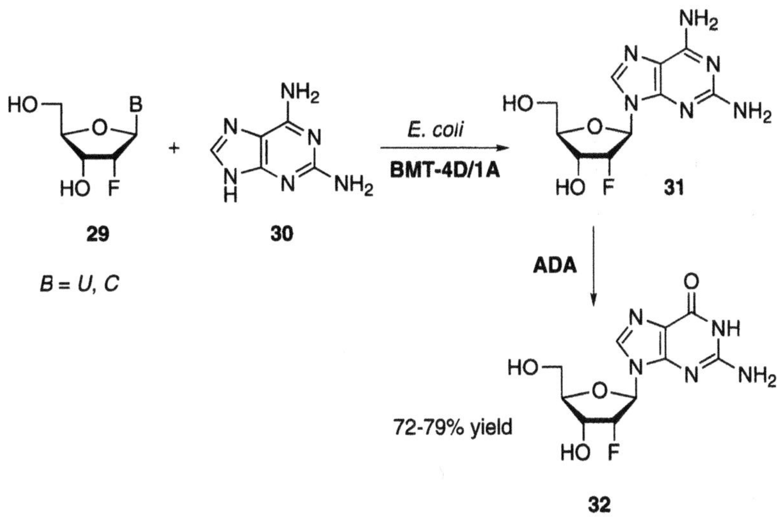

Scheme 8

Enzymatic Hydrolysis on the Sugar Moiety

Enzyme-catalyzed deacylations have been applied as a method for the selective removal of one or more acyl groups from polyacylated nucleosides. Early studies [20] have focused on the use of the dihydrocinnamoyl (*DHC*) group as an enzyme-labile protecting group removable by α-chymotrysin (α-CT). Although the enzyme showed an interesting tendency to attack preferably the 5′-position of uracil derivative **33** giving rise to protected nucleoside **34** (Scheme 9), this approach has not been further exploited.

Pseudomonas fluorescens lipase (PFL) and the protease subtilisin have been used to effect the regioselective deacylation of several 2′-deoxy-5-substituted uracil nucleosides (**36**) in phosphate buffer with *DMF* as cosolvent [21]. Thus, PFL preferably removes the acyl group on the secondary hydroxyl function leading to

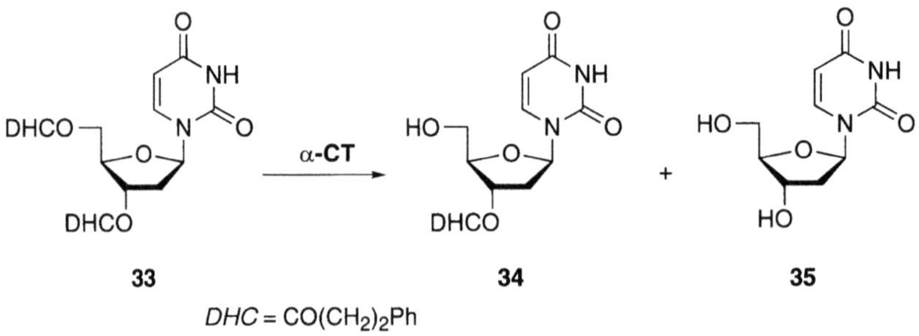

Scheme 9

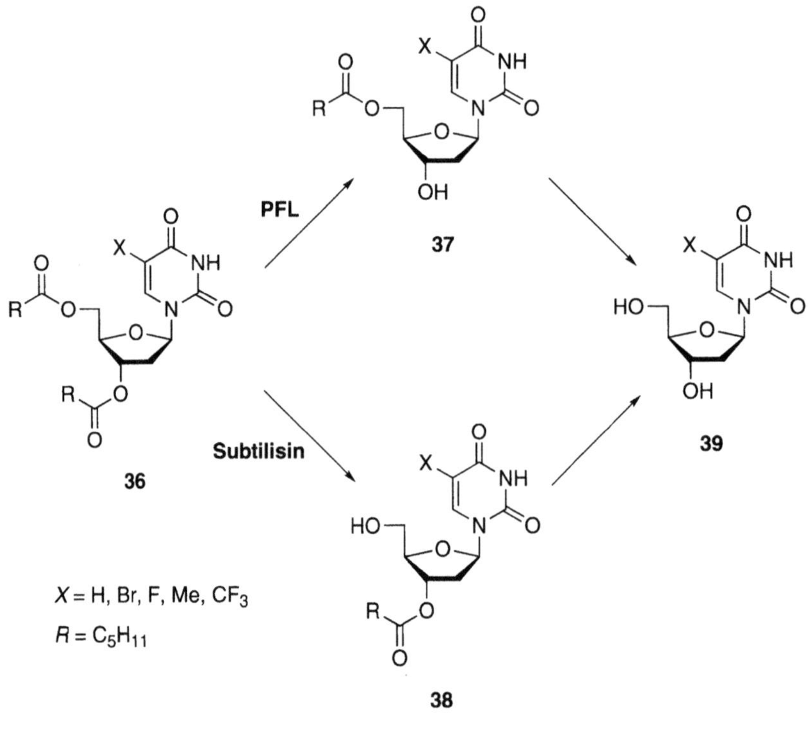

Scheme 10

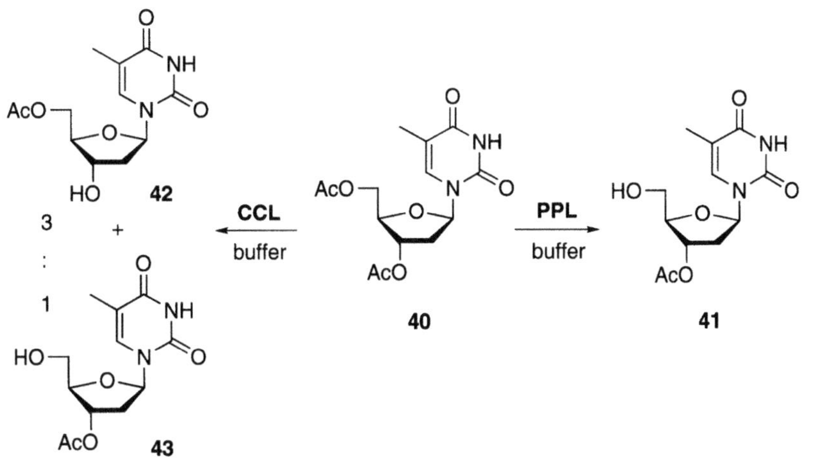

Scheme 11

the 5′-esters **37** in reasonably good yields. In contrast, subtilisin shows preference for 5′-acyl group cleavage, rendering the 3′-esters **38** (Scheme 10). With both enzymes, the completely deprotected nucleosides **39** have also been obtained.

Complementary results were obtained depending on the enzyme used [22]. Thus, when a lipase from porcine pancreas (PPL) in phosphate buffer was the catalyst for deacylation of 3′,5′-di-O-acetylthymidine (**40**), the 5′-O-acetyl group was selectively attacked leading to 3′-O-acetylthymidine (**41**) in very good yield (98%). In contrast, if *Candida cylindracea* lipase (CCL) was used in the catalysis, the 3′-ester (giving compound **42**) was hydrolyzed three times faster than the 5′-ester (giving compound **43**) (Scheme 11).

The regioselective hydrolysis of tri-O-acylated esters **44** has been carried out with pig liver esterase (PLE) in phosphate buffer with ethanol as cosolvent leading to 2′-O-monoacylated nucleosides **45** in high yield (Scheme 12) [23]. The markedly retarded rates of hydrolysis of the 2′-O-acyl esters of these arabinonucleosides suggest that they might function as slow-release lipophilic prodrugs with long serum lifetimes. However, they have greater aqueous solubility than the triesters and more stable pharmacokinetic properties than the fully deprotected arabinonucleosides. Similarly, it has been reported [24] that subtilisin acts on 9-(2′,3′,5′-tri-O-acyl-β-*D*-arabinofuranosyl)-adenosine affording the 3′,5′-diol derivative.

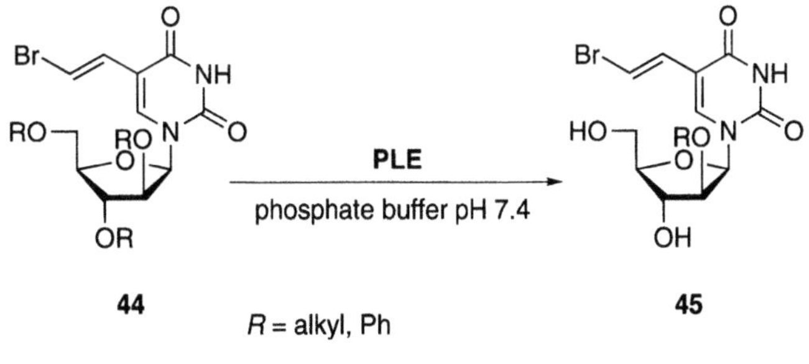

Scheme 12

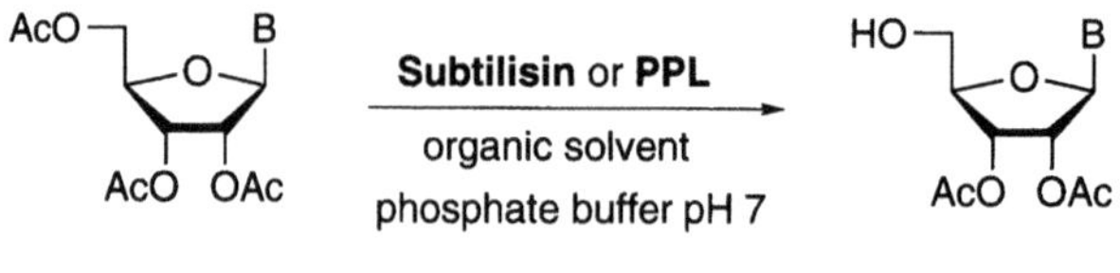

46 **47**

B = *U*, *C*, *A*, *G*, N-2-AcG, hypoxanthin-9-yl

Scheme 13

Subtilisin in phosphate buffer, with or without organic cosolvents (*DMF* or dioxane) selectively hydrolyzed the 5′-position of purine and pyrimidine tri-O-acylated esters **46** to give 2′,3′-di-O-acylribonucleosides **47** in 40–92% yield (Scheme 13) [25]. The least expensive crude lipase from porcine pancreas (PPL) also catalyzed the deacetylation, but resulted in poorer selectivity and a slower reaction rate.

The enzyme acetyl esterase (AE) from the flavedo of oranges, chemo- and regioselectively removes acetyl groups from purine nucleoside derivatives **17**, **18**, **20**, and **21** [16]. Thus, the adenosine- and guanosine derivatives **20** and **21**, which do not carry N-protecting groups in the nucleobases, are deprotected at the secondary 3′-OH to give nucleoside derivatives **49** and **51**, respectively (Scheme 14). However, if the amino groups present in the purine bases are masked as phenylacetamides, as with compounds **17** and **18**, the regioselectivity displayed by the enzyme is reversed, and the primary 5′-OH groups of the nucleosides are liberated to give base-protected nucleosides **48** and **50**, respectively.

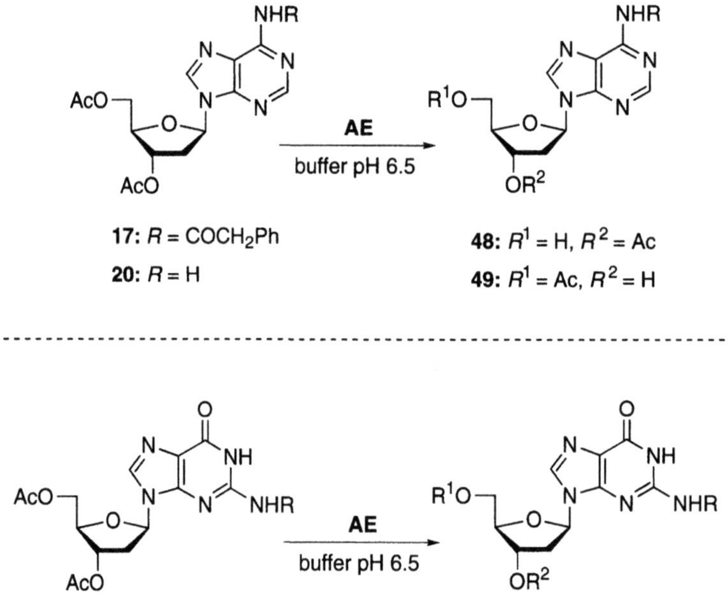

Scheme 14

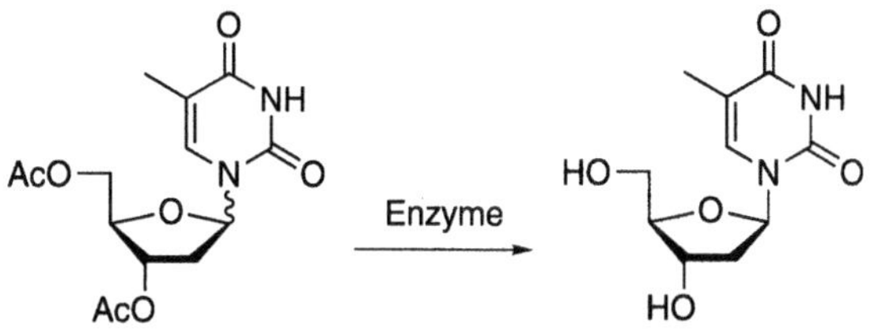

52 53

WGL	29% yield (pure β)
PLE	31% yield (pure β)

Scheme 15

Wengel et al. [26] have reported the first attempt to use biotransformations to solve the basic problem of anomer separation in nucleoside chemistry when a convergent strategy is used. Thus, lipase-catalysed deacylations of an anomeric mixture of peracetylated 2′-deoxyribofuranosyl thymine (**52**) are shown in Scheme 15. Generally, the diastereoselectivity was more pronounced in pure phosphate buffer than in phosphate buffer containing 10% of *DMF*. Wheat germ lipase (WGL) and porcine liver esterase (PLE) catalyzed diastereoselective the deacetylation of **52** affording the pure β-anomer thymidine (**53**β) as the only completely deprotected nucleoside product.

Enzymatic Protection of Nucleosides by Acylations

Selective modification of nucleosides with several functional groups of very similar chemical reactivities is an interesting subject of study and a fundamental challenge to organic chemists. Thus, *Klibanov et al.* [27] have studied the catalytic activity of the protease subtilisin, using an activated ester like trichloroethyl butyrate to avoid the reversibility of the process, in the regioselective esterification of adenosine (**54**) and uridine (**55**) in anhydrous *DMF*. The regioselectivity of the enzymatic transesterification is not that good, but the hydroxyl group at the 5′-position of nucleosides **54** and **55** was preferentially acylated (Scheme 16). An important fact was that the purity of subtilisin is not essential for either the yield or the regioselectivity of the enzymatic acylation.

Acylation of 2′-deoxy-5-substituted uridine nucleosides **39** with acid anhydrides in the presence of PFL in organic solvents [28] showed only moderate

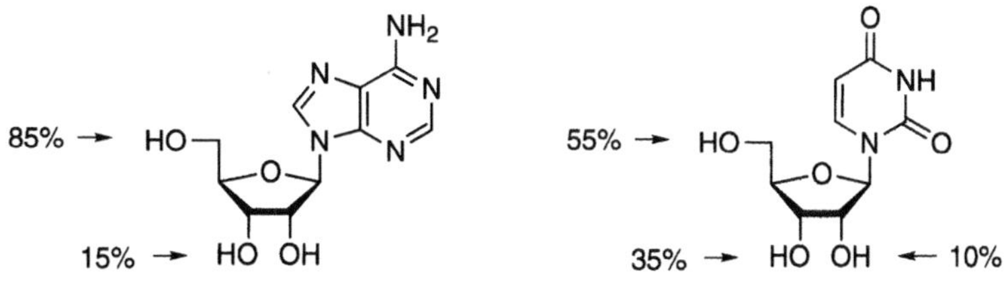

54 55

Scheme 16

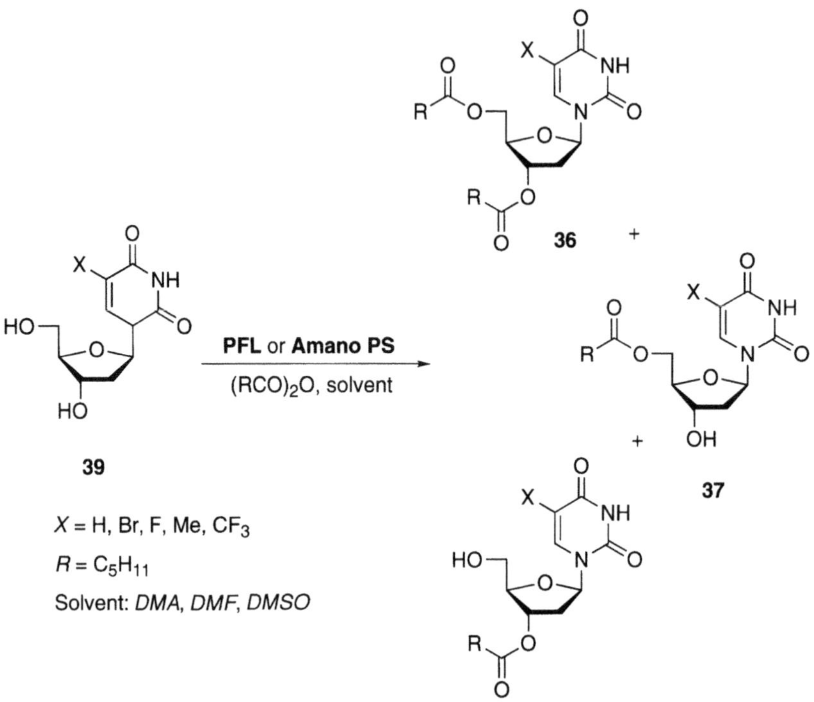

Scheme 17

selectivity towards the 5′-hydroxyl group, giving rise to derivatives **37** in addition to diprotected compound **36**. When Amano PS was used instead [29], the 3′-hydroxyl function was acylated almost exclusively yielding nucleoside derivative **38** (Scheme 17).

Highly regioselective acetyl transfer reactions to the carbohydrate moiety of nucleosides have been carried out by employing a subtilisin mutant [30] (subtilisin 8350, or even the much more stable subtilisin 8397 [31]) obtained *via* site-specific mutations in anhydrous *DMF*. This mutant enzyme transfers the acetyl group from isopropenyl acetate to the primary hydroxyl groups of various ribonucleosides and 2′-deoxyribonucleosides **56** giving the 5′-O-acylated derivatives **57** in high yields (Scheme 18). The high regioselectivity of the mutant enzyme was attributed to the fact that it binds the reaction transition state more strongly than does the parent enzyme.

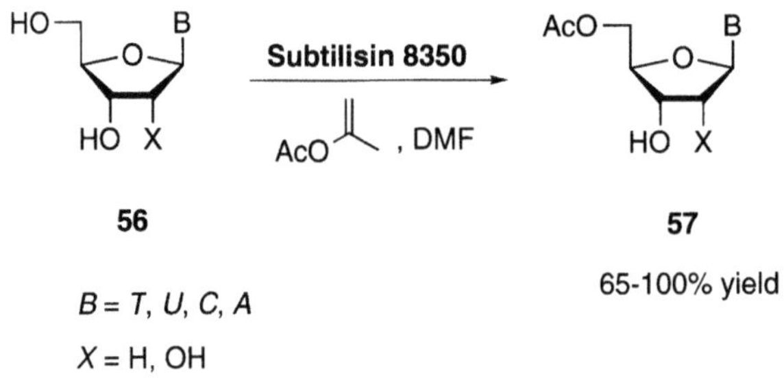

Scheme 18

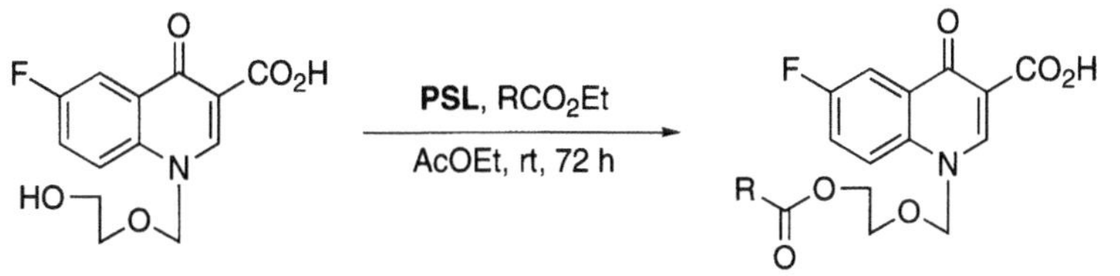

58

59

60-87% yield

R = Me, Et, ^{n}Pr, ^{s}Bu, CH_2=CH, MeCH=CH, MeCHCl

Scheme 19

The reactivity and selectivity of subtilisin was later increased by changing the acylating agent and the solvent from trichloroethyl butyrate in *DMF* to trifluoroethyl butyrate in anhydrous pyridine, thus affording 5′-O-acylated derivatives **57** (uridine: N-4-anisoylcytidine adenosine N-6-benzoyladenosine) in 67–82% yield [32].

As a part of a program to design new regioselective enzymatic transformations of nucleosides, *Gotor et al.* have started with the lipase-mediated acylation of acyclonucleosides (Scheme 19) [33]. Thus, acyclonucleosides **59** bearing different substituents in the 2-hydroxyethoxymethyl chain were prepared in almost quantitative yields using *Pseudomonas cepacea* lipase (PSL) and several ester derivatives through a transesterification reaction from precursors **58**. This was the first example of the use of enzymes in acyclonucleoside chemistry that allowed a facile preparation of prodrugs **59** of 4-quinolone acyclonucleosides which were tested as antiviral agents against a wide variety of assay systems [34].

In addition, the same authors have used oxime esters as irreversible acyl transfer agents in the regioselective acylation of 2′-deoxynucleosides **60** on the secondary hydroxyl group (Scheme 20) instead of the chemically more reactive primary alcohol using PSL in pyridine [35]. This procedure is very versatile because acylation with various oxime esters bearing saturated or unsaturated chains, yields exclusively the 3′-O-acyl derivatives **61**. However, N-acylation of 2′-deoxyadenosine is not observed. Similarly, *Candida antarctica* lipase (CAL) showed high regioselectivity toward the primary hydroxyl group of both deoxy- (**60**) and ribonucleosides (**63**) [36]. 2′-Deoxynucleosides such as thymidine and 2′-deoxyadenosine, (**60**) were acylated with oxime esters carrying saturated, unsaturated, aromatic, and functionalized chains, giving 5′-O-acylated compounds (**62**) together with small quantities of the 3′-O-acylated regioisomer **61**. Uridine, adenosine, and inosine as representative ribonucleosides (**63**) were acylated exclusively at the 5′-hydroxyl group (**64**) by using the same methodology. Furthermore, these authors have studied in depth several factors that affect this kind of enzymatic process, such as the alcohol inhibition and specificity of CAL in organic solvents [37], as well as solvent effects on equilibrium and individual rate constants [38].

To decrease toxicities and improve activities, chemical modifications of nucleosides, including protection, have been studied. Thus, *Ozaki et al.* [39] have used enzymatic transesterification reactions to prepare acyl-protected derivatives of

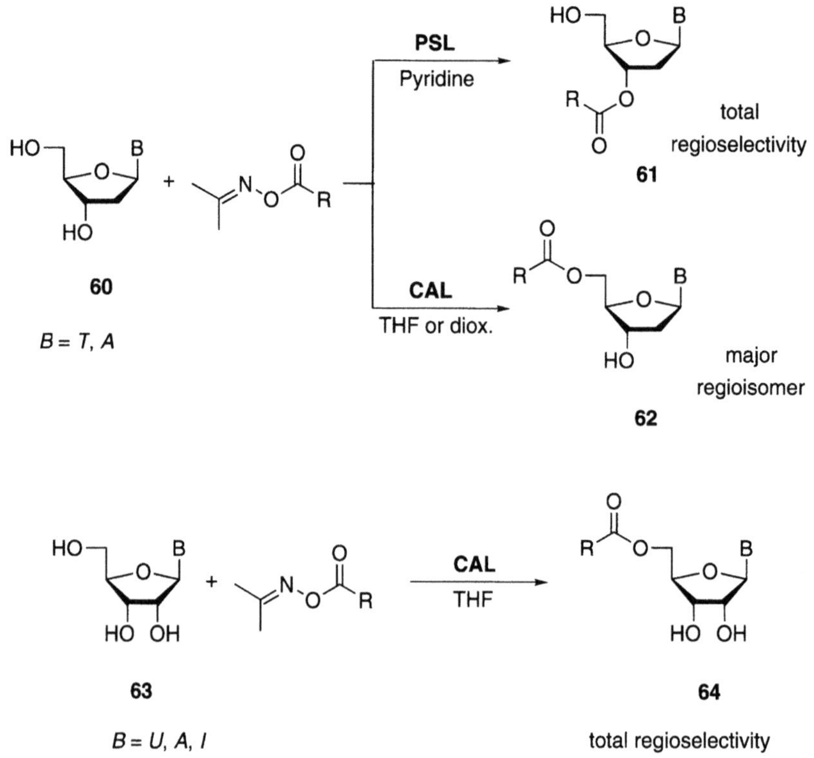

R = Me, ^{n}Pr, $Me(CH_2)_6$, $Me(CH_2)_8$, CH_2=CH, MeCH=CH, Ph, $ClCH_2$

Scheme 20

5-fluorouridine (**65,** Scheme 21). They used anhydrides (*n*-octanoic anhydride showed the highest selectivity) and lipase PS (a lipase from *Pseudomonas sp.*, Amano) or lipase KWI-56 (a lipase from *Pseudomonas sp.*, Kurita water industries Ltd.) to almost exclusively acylate the 3′-position in this nucleoside derivative **65** (isolated yields in parentheses). Additionally, lipase M (from *Mucor javanicus*, Amano) acylated the 2′-hydroxyl group. However, CAL acylated the 5′-OH to give the corresponding derivative in excellent yield. Although several solvents were checked, *THF* was found to be suited best for this reaction. Furthermore, selective protection of uridine **55** and arabinosyluracil **66** was also investigated. In general, the acylation of these compounds proceeded faster than the acylation of **65** under the same conditions with *n*-caproic anhydride. As shown for the fluoroderivative **65**, the 3′-OH of uridine was acylated by lipase PS and KWI-56, and the 5′-OH was acylated by CAL. The 5′-OH of arabinosyluracil was selectively protected by CAL.

2′,3′-Dideoxynucleosides have been introduced as chemotherapeutic agents with anti-retroviral activity against HIV. Among these, 2′,3′-dideoxyinosine (DDI, **74**) has been approved for alternative treatment to AZT. *Santaniello et al.* [40] have described a chemoenzymatic synthesis of DDI based on enzymatic acylation studies of inosine and 2′-deoxyinosine. Thus, the selective acylation of the hydroxyl groups of the nucleosides inosine and 2′-deoxyinosine **67** has been achieved in presence of CAL in organic solvents (Scheme 22). Also, the enzymatic

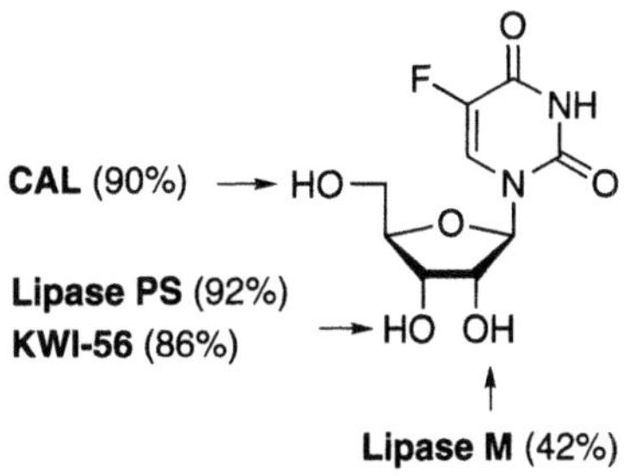

65

^{n}Octanoic anhydride

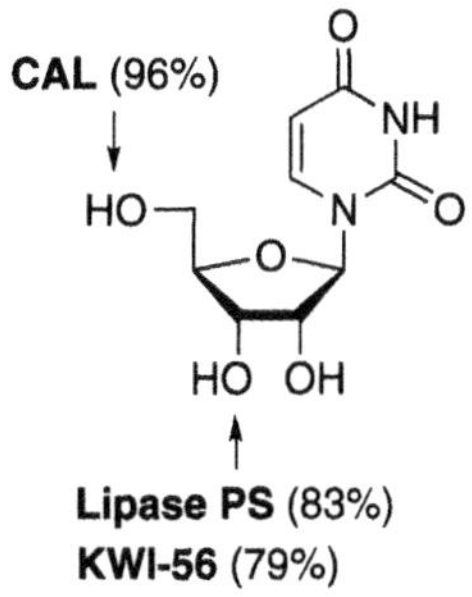

55

^{n}Caproic anhydride

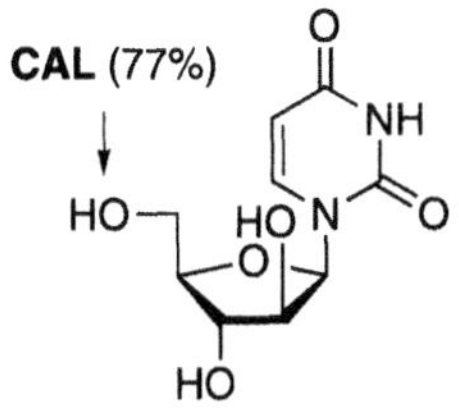

66

^{n}Caproic anhydride

Scheme 21

hydrolysis of the triacetate and diacetate **69** in presence of CAL in aqueous buffer at $pH=7$ or in water-saturated chloroform was performed and was found to be complementary to the lipase-catalysed acylation in organic solvents. With these results in hand, they developped a new chemoenzymatic synthesis of 2′,3′-dideoxyinosine **74** starting from the 5′-acetate **73** enzymatically prepared by the transacetylation procedure. Starting material **72** which is relatively expensive, was prepared from readily available 2′-deoxyadenosine (**71**) by a deamination reaction efficiently catalyzed by adenosine deaminase (ADA). The overall sequence for the chemoenzymatic synthesis of DDI is depicted in Scheme 22.

Enzymatic Protection of Nucleosides by Alkoxycarbonylations

An important and synthetically relevant transformation in nucleoside chemistry is the selective alkoxycarbonylation of the sugar moiety in order to obtain nucleoside carbonates, which play an important role in the synthesis of oligonucleotides, and other derivatives assayed in medicine, such as dinucleoside carbonates. Scheme 23 summarizes a general, new, and simple procedure for the synthesis of pyrimidine and purine 3′-carbonates **75** from 2′-deoxynucleosides **60** using PSL and O-alkoxycarbonyl oxime [41]. In this method, no previous protection of the primary hydroxyl group is necessary, as has been traditionally described for the preparation of these compounds. 5′-O-Carbonates **76** and **77** of ribonucleosides and 2′-deoxyribonucleosides could be obtained by enzymatic alkoxycarbonylation with

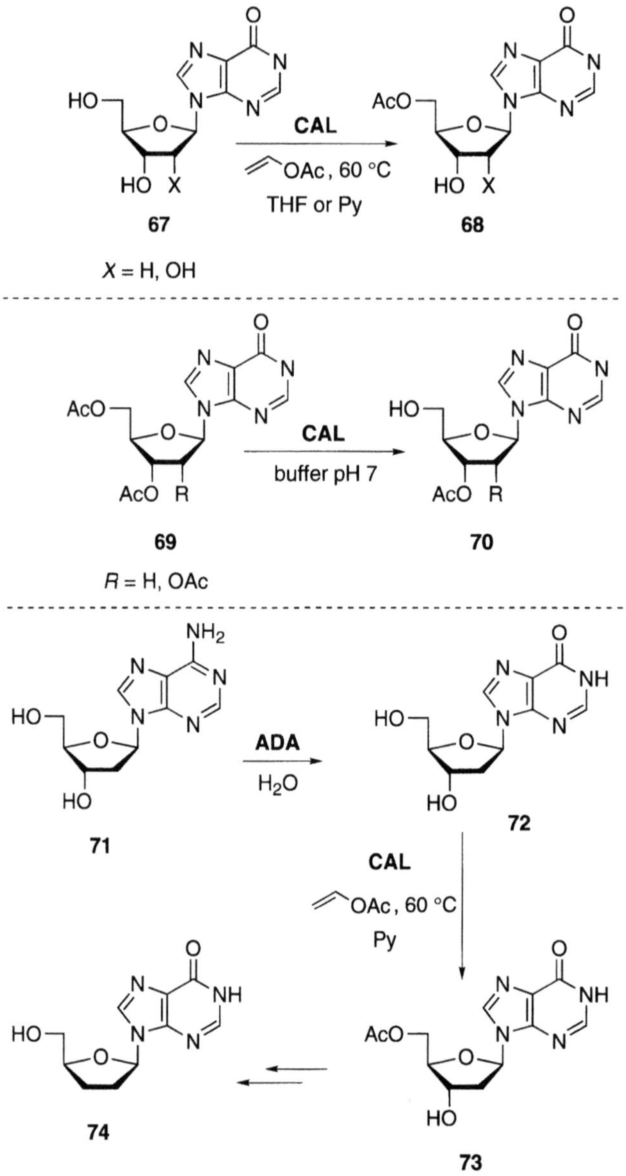

Scheme 22

CAL and oxime carbonates [42], the latter being easily prepared from chloroformates. Ribonucleosides **63** yield two kinds of 5′-O-carbonates **77**, depending on whether the alkoxy or the acetone oxime moiety acted as the leaving group. In the case of 2′-deoxynucleosides, the leaving group was always the acetonoxime moiety, giving rise to the regioselective formation of the corresponding 5′-O-alkyl carbonates together with small amounts of the 3′-O-regioisomer and diacylated compounds.

One of the most widespread groups in chemotherapy is the carbamate moiety which occurs, among others, in several classes of antitumorous products, and is also used for the increase of permeation through biological membranes. Bearing

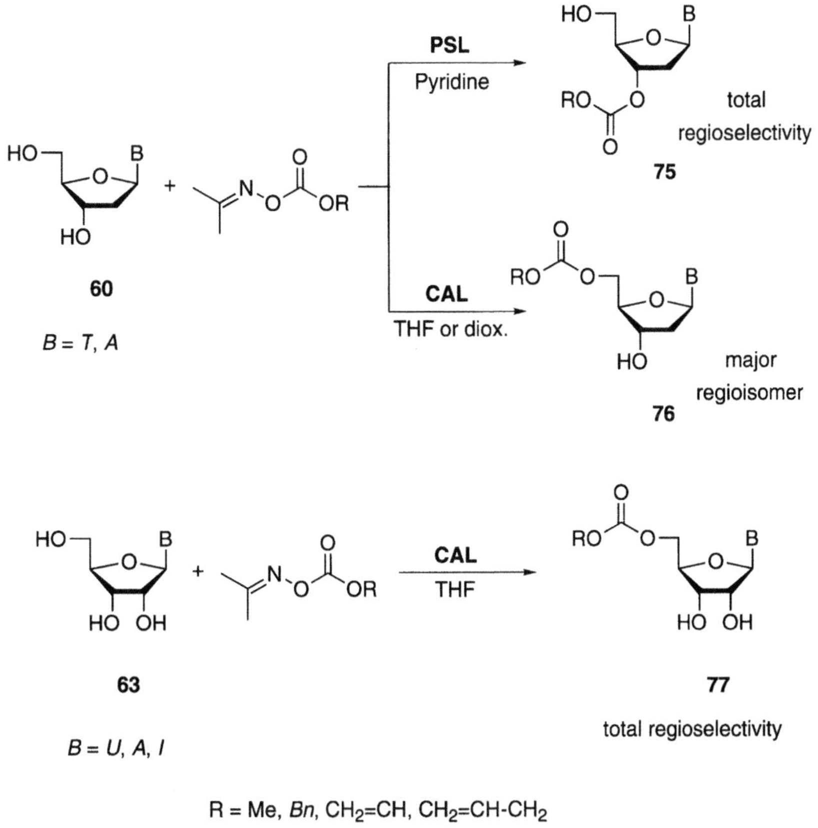

Scheme 23

this in mind, and taking into account the interesting features shown by the carbamate moiety in the nucleoside field, such as phosphate simulation ability and chemical and enzymatic stability, *Gotor et al.* were interested in proposing a pathway which would allow them to synthesize these nucleoside carbamates through a simple, generic procedure. Testing of various lipases revealed them to be unable to catalyze the reaction between vinyl carbamates and nucleosides. One feasible explanation might be that some carbamates have been shown to be good inhibitors of many serine hydrolases. Taking this into account, it was thought appropriate to plan the synthesis in a two-step procedure (Scheme 24) [43]. The key step is the enzymatic synthesis of 5′-O-vinyloxycarbonylnucleosides **79** or 3′-O-vinyloxycarbonylnucleosides **80** in a regioselective way using CAL or PSL catalysis, respectively. In a second aminolysis step these carbonates yield the corresponding urethanes **81** and **82**. This methodology allows the assembly of ammonia, amines, amino alcohols, and *L*-amino acids to 2′-deoxynucleosides (3′- or 5′-positions) or ribonucleosides (5′-position) [44].

In view of the previous results, it is believed that nucleosides with different orientations of the hydroxyl groups (**83–88**) could be an interesting subject of study with respect to the preference of some lipases according to the geometry of the substrate employed [45]. Thus, 5′-O-acyl and 5′-O-alkoxycarbonyl derivatives of α-, anhydro-, xylo-, and arabino-nucleosides are obtained through a lipase-

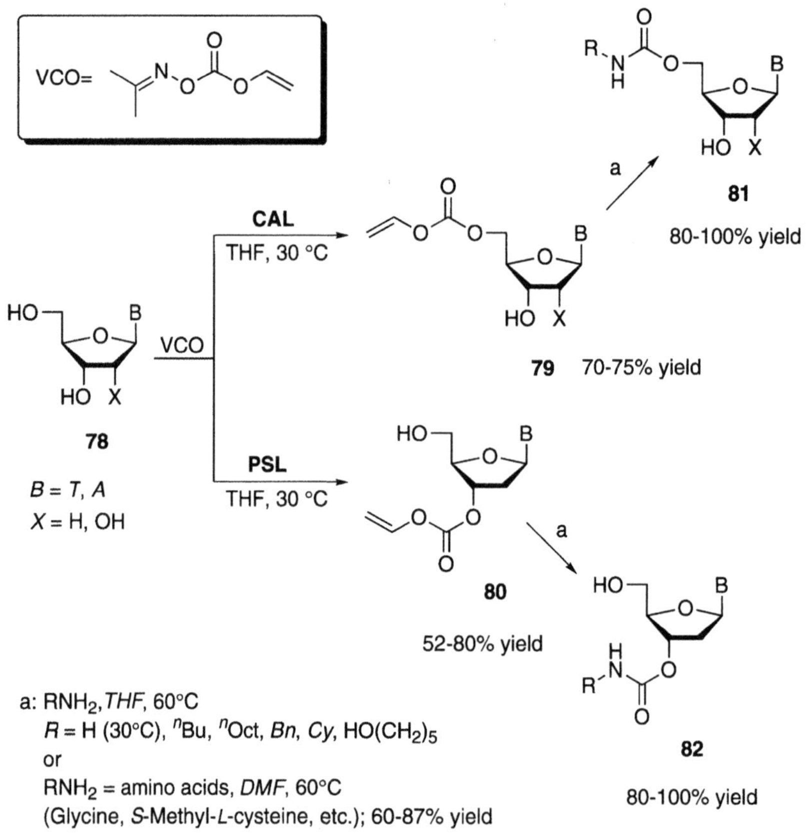

Scheme 24

mediated reaction with CAL by using acetoxime butyrate or butyric anhydride, together with benzyloxycarbonyl-O-acetoxime as acylating or alkoxycarbonylating agents (Scheme 25).

For the synthesis of the highly hindered 3′-O-acetyl α-thymidine **91** and xylo-thymidine **94** [46], the same authors prepared the 5′-*Cbz*-derivatives **89** and **92** as previously described, and then chemically carried out an acylation reaction in the 3′-OH free hydroxyl group giving rise to fully protected nucleosides **90** and **93**, since benzyloxycarbonyl group could be easily removed (Scheme 26).

Aminosugar nucleosides are known to possess strong antibacterial, anticancer, and biosynthetic inhibitory properties. Therefore, considerable effort has been devoted to the preparation of this kind of compounds. Most of the procedures depend strongly on the nature of the nitrogenated base present in the starting nucleoside. Schemes 27 and 28 show novel and general chemoenzymatic procedures to obtain the 3′-amino-xylo-nucleosides [47] in both 2′-deoxynucleosides and ribonucleosides. The synthetic scheme is based on the 5′-directed intramolecular nucleophilic substitution at the 3′-activated position of the nucleoside. The approach of the incoming group to this position takes place both regio- and stereoselectively from the most hindered face of the nucleoside **96** or **101/104**. The methodology presented is applicable to 2′-deoxyribonucleosides (Scheme 27) and ribonucleosides (Scheme 28), regardless of their nitrogenated base.

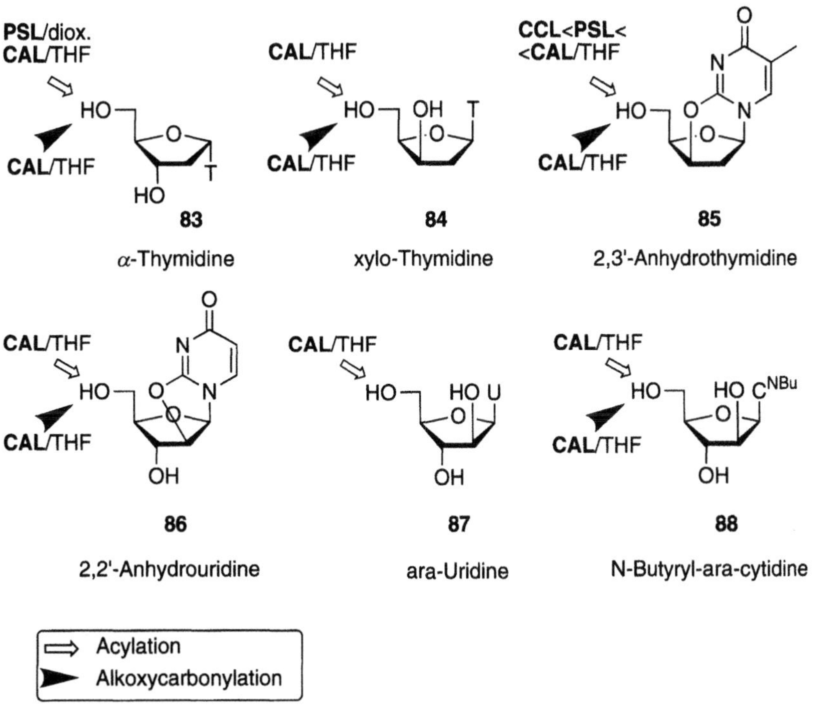

Scheme 25

A chemoenzymatic procedure for the synthesis of 3′- and 5′-carbazoyl nucleoside derivatives prepared for the first time is shown in Scheme 29 [48]. This process involves the regioselective enzymatic alkoxycarbonylation of nucleosides **106** and the subsequent transformation with hydrazine into novel carbazoyl nucleoside derivatives **107** or **108**. Taking into account previously reported data, 3′-alkylidencarbazoyl-2′-deoxynucleosides (**110**) as well as 5′-alkylidencarbazoyl-2′-deoxynucleosides and 5′-alkylidencarbazoyl ribonucleosides (**109**) emerge as interesting targets combining structural features found in both therapeutic nucleoside derivatives and fungicide/herbicide nucleoside analogues.

Modulation of gene expression by antisense technologies requires the development of modified oligonucleotides possessing enhanced cellular uptake,

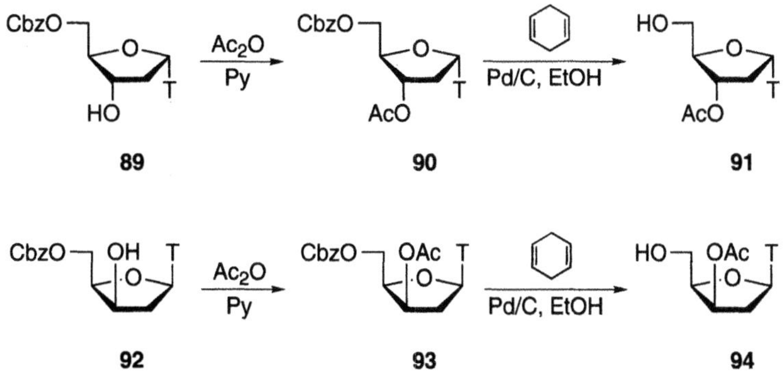

Scheme 26

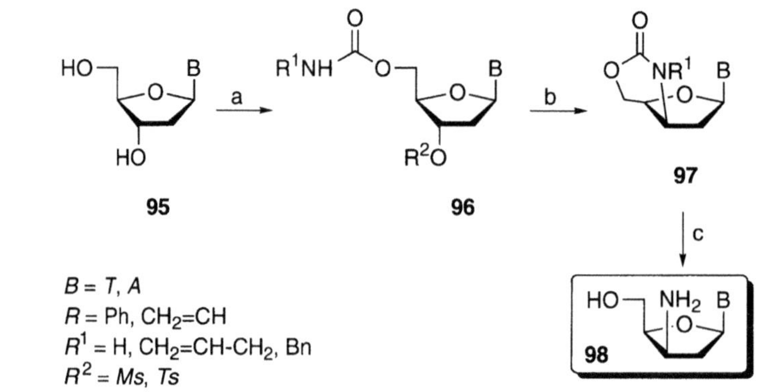

a: *i* : $ROCO_2N{=}CMe_2$, CAL, *THF*; *ii* : R^1NH_2, *THF*; *iii* : R^2Cl, *Py*;
b: NaH, *THF*; c: *i* : LiOH, $EtOH\text{-}H_2O$; *ii* : Pd, HCO_2H, MeOH

Scheme 27

resistance toward degradation by nucleases, and appropriate hybridization to target natural oligonucleotides. Consequently, these modified oligonucleotides are now being actively investigated as a new generation of pharmaceuticals. In the last decade, a great deal of effort has been directed toward the synthesis of analogues with an altered phosphodiester linkage, and one of the most important modifications is the complete substitution of the phosphate internucleoside bridge. Thus, several properties of the natural oligonucleotides have been improved for the

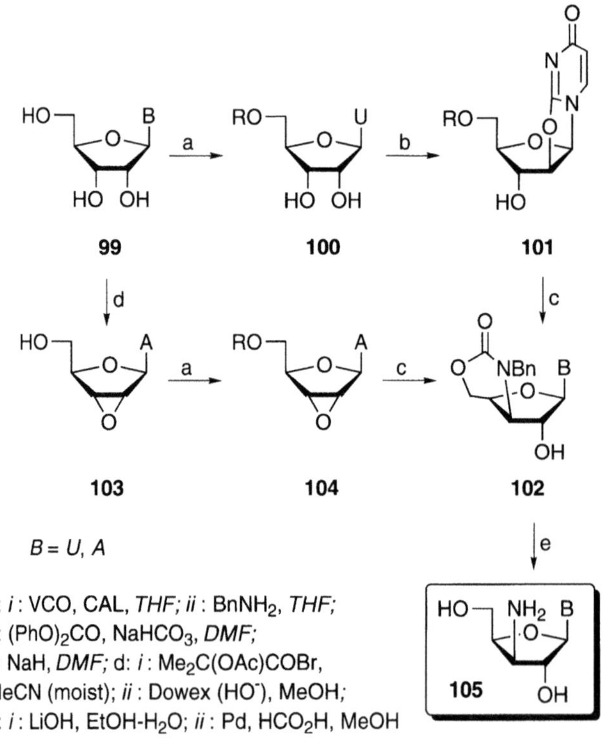

a: *i* : VCO, CAL, *THF*; *ii* : $BnNH_2$, *THF*;
b: $(PhO)_2CO$, $NaHCO_3$, *DMF*;
c: NaH, *DMF*; d: *i* : $Me_2C(OAc)COBr$,
MeCN (moist); *ii* : Dowex (HO^-), MeOH;
e: *i* : LiOH, $EtOH\text{-}H_2O$; *ii* : Pd, HCO_2H, MeOH

Scheme 28

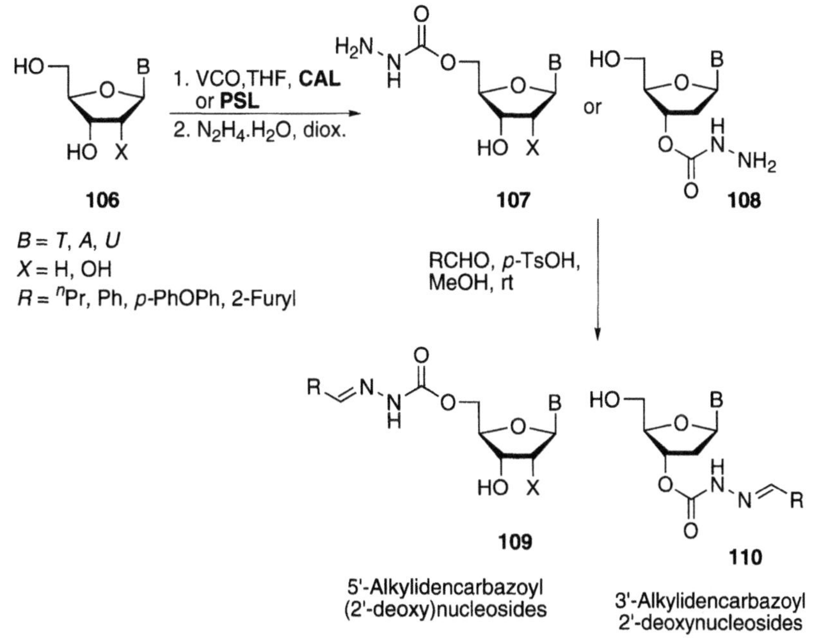

Scheme 29

potential therapeutic application of the antisense strategy. Up to date there has been no investigation of backbone linkages that contain carbazoyl groups, and yet this type of linkage appears to have properties that make it an attractive surrogate for the phosphodiester linkage, being non-ionic, hydrolytically stable, and non-chiral.

The synthesis of backbone modified dinucleotide analogues **112** is shown in Scheme 30 [49], in which the natural phosphodiester linkage is replaced by a 3′–5′ carbazoyl linkage. The bridge was formed through a coupling reaction between an appropriate 3′-carbazoyl nucleoside analogue **108** and an aldehyde nucleoside derivative **111**. It is noteworthy that the starting nucleosides could be common materials to obtain the 3′-carbazoyl nucleoside derivatives by means of a simple previously described chemoenzymatic procedure, the aldehyde nucleoside could be obtained by an oxidation reaction.

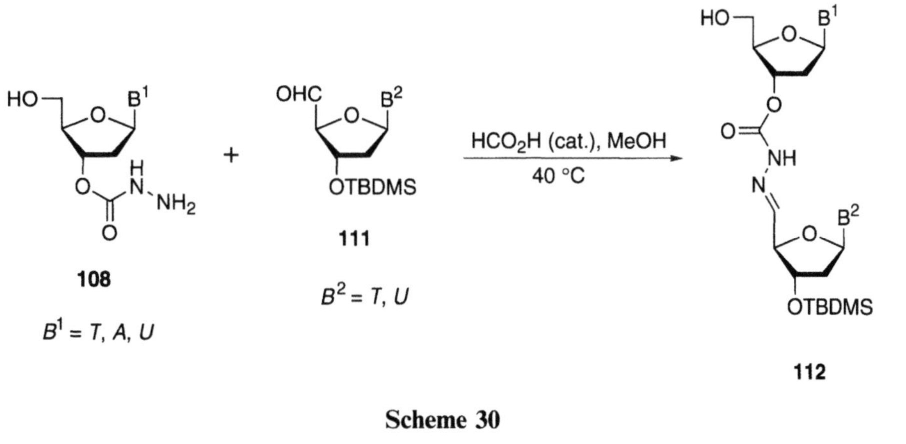

Scheme 30

Gotor et al. have studied these lipase-catalyzed alkoxycarbonylation processes in depth from the point of view of the influence of the carbonate on the regioselectivity [50] in order to understand the behaviour of the enzymes.

Enzymatic Protections by *β*-Galactosidase

The synthesis of various galactose containing disaccharide nucleosides **115** has been achieved by utilizing the transglycosilation potential of *β*-galactosidase from *Aspergillus oryzae* [51] (Scheme 31). Thus, using *p*-nitrophenyl-*β*-*D*-galactoside **113** as galactosyl donor, 2-deoxyuridine, uridine, thymidine, and adenosine have proven to be useful acceptors for enzyme-catalyzed disaccharide nucleoside formation. The regiochemistry of the products **115** showed that only 5′-protected derivatives were formed.

Nucleoside Enzymatic Resolutions

Ohno et al. [52] have reported an efficient synthesis of the nucleoside (–)-cordycepin (**120**) in an enantioselective and stereocontrolled manner by chemo-

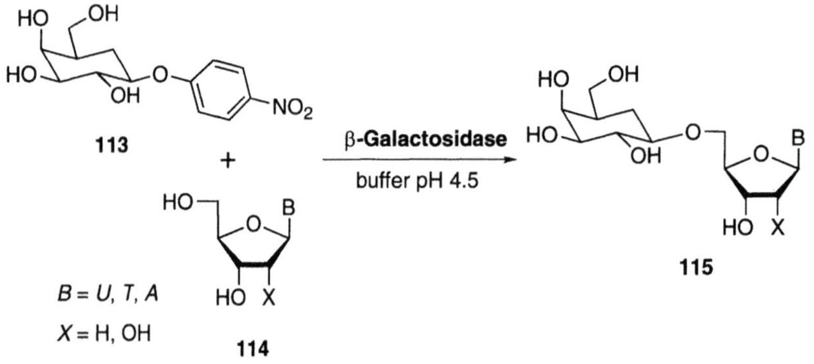

Scheme 31

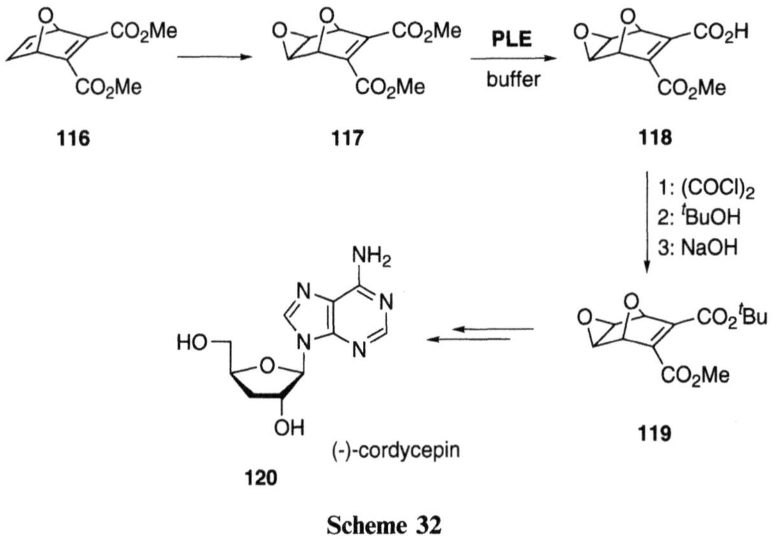

Scheme 32

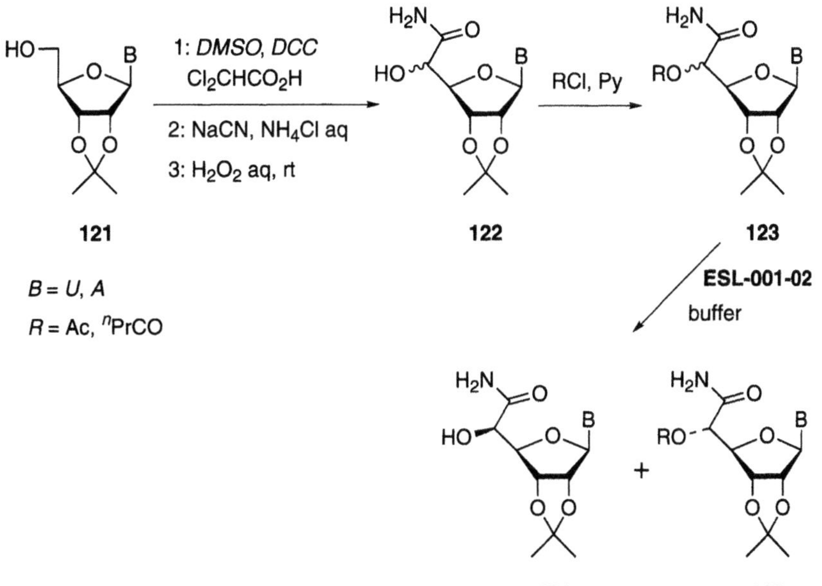

Scheme 33

enzymatic strategy, starting from the *Diels-Alder* adduct of furan and dimethyl acetylenedicarboxylate **116** (Scheme 32). The symmetric unsaturated dimethyl ester **117** was almost quantitatively hydrolyzed with pig liver esterase (PLE) to yield half-ester **118** in reasonably high optical yield. After the enantiomer conversion (from **118** to **119**), (–)-codycepin (**120**) was obtained from **119**. This is one of the first synthetic approaches to the sugar moiety of nucleosides starting from non-carbohydrate synthons.

Wang et al. [53] have reported a one pot preparation of a mixture of α-*L*-taluronamide **124** and β-*D*-alluronamide **125** nucleoside derivatives (Scheme 33). Such derivatives can be considered as intermediates to uracil polyoxin C, capuramycin, and other nucleoside antibiotics. Nucleoside derivatives **124** and **125** were prepared *via* a one-pot reaction sequence of oxidation, nucleophilic addition, and hydrolysis of **121** to give hydroxyamide **122**. The resulting diastereoisomers, which are difficult to separate by conventional chromatography, were resolved by stereoselective deacylation of their 5′-O-acylated derivatives **123** with thermophilic enzymes (from esterase/lipase CloneZyme (ESL-001) library). Themophilic enzyme ESL-001-02 exhibited good selectivity in the hydrolysis of (*R*)-C5′ substrates.

Enzymatic elaboration of the side chain in nikkomycin B

A stereoselective synthesis of the versatile chiral synthon (2*S*,3*S*)-**127** possessing two stereogenic centres was achieved using the chemoenzymatic method [54], starting from *p*-hydroxybenzaldehyde (**126**). The enzymatic resolution of the racemic mixture of intermediate **127** using lipase Amano PS gave enantiomerically pure substrate (2*S*,3*S*)-**127** (Scheme 34). The conversion of the latter into the homochiral intermediates (2*S*,3*S*,4*S*)-**129** and (2*S*,3*S*,4*S*)-**130**, corresponding to the N-terminal amino acid moiety of nikkomycin B, and their reaction with **131**, the

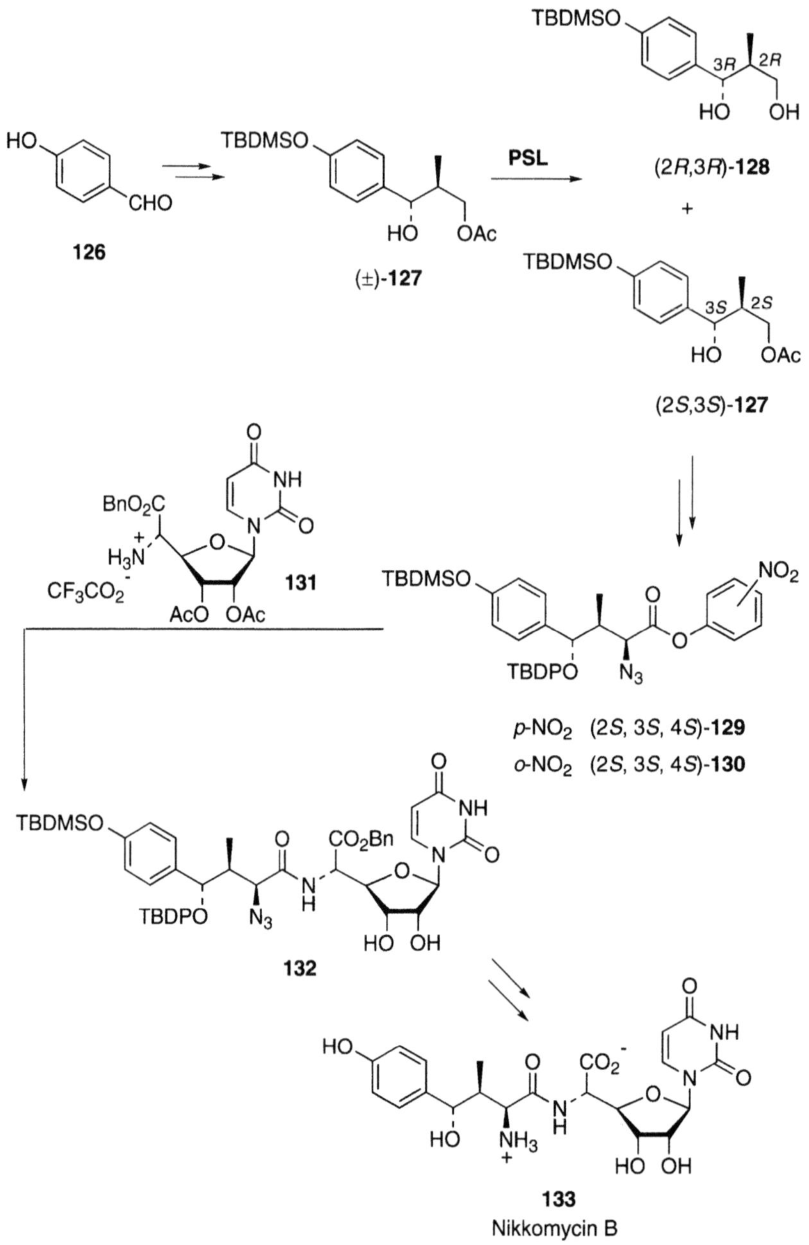

Scheme 34

corresponding part of the C-terminal nucleoside amino acid, afforded the important intermediate **132** for the synthesis of nikkomiycin B (**133**).

Enzymatic Protecting Strategies in Nucleopeptides

Aminoacyl nucleosides (**134–136**) and their oligonucleotide derivatives are compounds of high significance in biological chemistry, since it has been established that the 3′-terminus of all *t*-RNAs contains the common CCA sequence

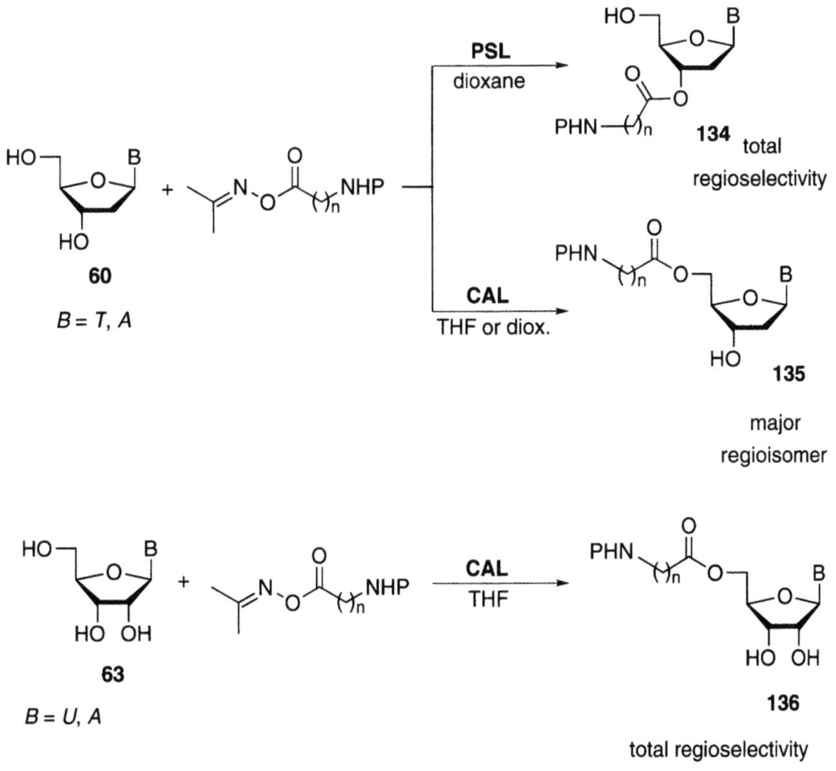

Gly (*n* = 1), *β-Ala* (*n* = 2); *P* = *Cbz*, *Boc*

Scheme 35

and the 3′-terminal adenosine *cis*-diol system is a site of attachment of the amino acid. Therefore, aminoacylated derivatives of oligoribonucleotides are important tools in the study of the specific incorporation of amino acid into proteins. *Gotor et al.* have applied the methodology previously described to obtain aminoacylated nucleosides in a regioselective manner by using the lipases PSL and CAL and employing N-protected acetoxime aminoacyl esters as acylating agents (Scheme 35) [55]. It is noteworthy that, in contrast to other methods reported, the exocyclic amino function in adenine nucleosides remains unaffected. Amino acids protected with benzyloxycarbonyl moieties are better substrates than those derivatized as *tert*-butyloxycarbonyl groups. With respect to α-branched amino acids, neither *L* nor *D* seems to fit the steric requirements of the enzyme's active sites.

Acid- and base-labile multifunctional nucleopeptides have been selectively constructed under mild conditions by means of enzymatic protecting group techniques [56]. The nucleopeptides **137** and **140** have the nucleobases masked with enzyme-labile phenylacetamide groups and the carboxy groups of the amino acids masked as methyl or *MEE* esters (Scheme 36). From the nucleoamino acid esters **137** the C-terminal protecting groups are selectively removed with the protease papain from *Carica papaya* in a cysteine buffer at *pH* 6.6 and 37°C giving nucleopeptide **138**. Under these conditions the protease did not attack the O-acetates. However, the nucleobase in **137** is selectively N-deprotected by means of

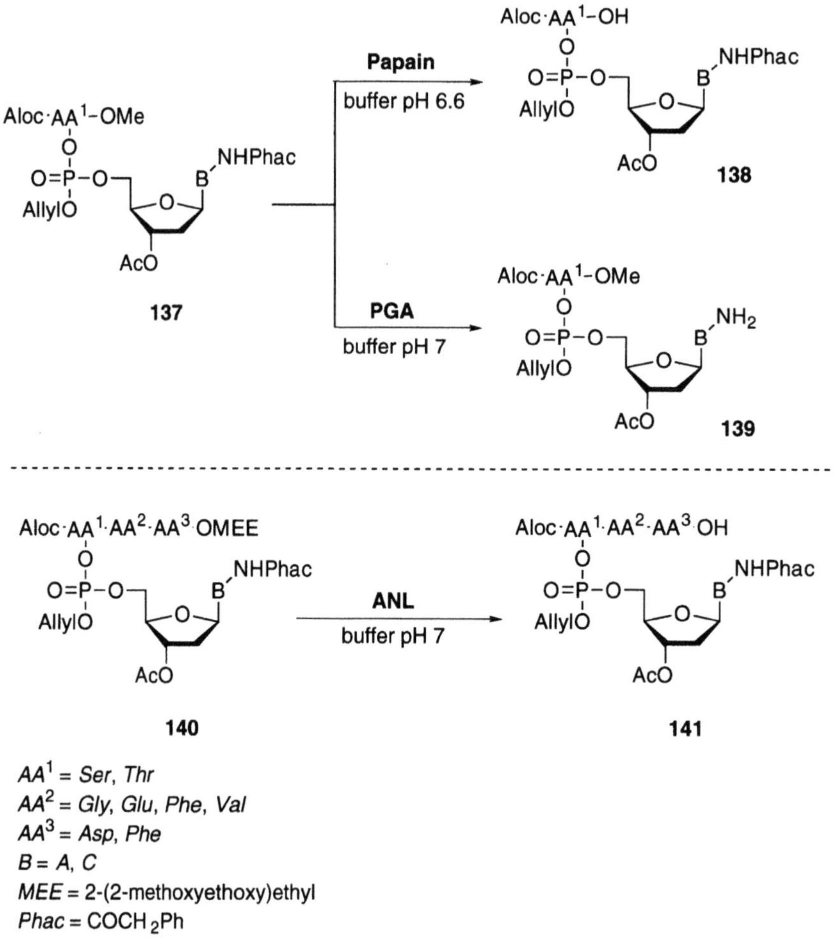

Scheme 36

penicillin G acylase (PGA) in phosphate buffer ($pH = 7$) MeOH (80:20) at room temperature, giving rise to compound **139**. Upon treatment of the nucleopeptide acid *MEE* esters **140** with lipase from *Aspergillus niger* (ANL) in analogously phosphate buffer ($pH = 7$) acetone (90:10) a 37°C, the C-terminal carboxylic acid was smoothly deprotected giving derivative **141**. The biocatalyst tolerated the presence of purine and pyrimidine bases and different amino acids and amino acid sequences in the nucleopeptides.

This strategy was later complemented by the same authors [57] for the hydrolysis of the 3′-O-acetyl group in nucleopeptides **137** using wheat germ lipase (WGL) in phosphate buffer at $pH = 6.5$ to give exclusively nucleopeptide **142** (Scheme 37). However, they show that the enzyme butyrylcholine esterase (BCE) from horse serum was advantageously employed to smoothly deprotect C-terminal carboxylic acids like **143** to give compound **141**. Thus, even on these multifunctional, complex conjugates, the mild enzymatic transformations occurred without any undesired side reactions.

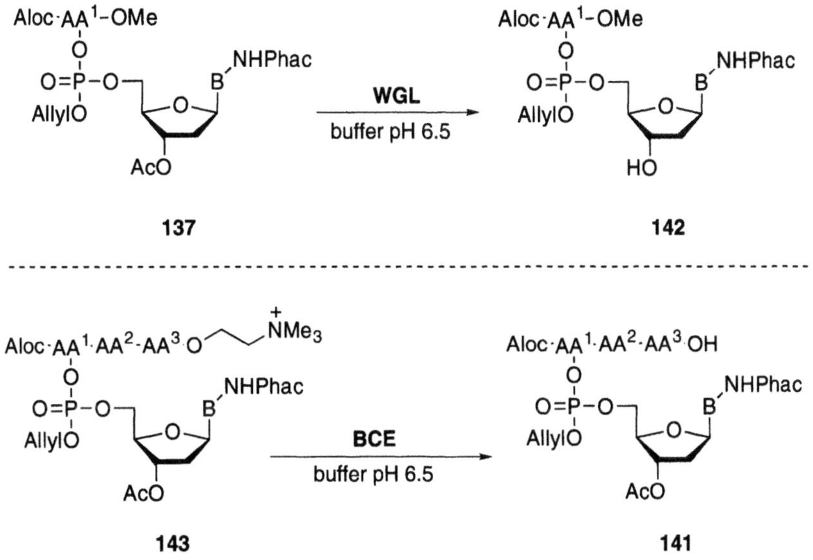

Scheme 37

Other Enzymatic Transformations

The cyclase of *Aplysia californica* [58] catalyzed an alternative mode of cyclization of nicotinamide guanine dinucleotide **144** and nicotinamide hypoxantine dinucleotide **145** to form the cyclic GDP-ribose **146** and cyclic HDP-ribose **147**, respectively (Scheme 38). In these cyclic nucleotides, the newly formed glycosyl bonds are attached onto the N-7 nitrogen of the purine rings instead of the N-1 nitrogen as in cyclic *ADP*-ribose. In addition, a series of NMR techniques were employed to deduce its chemical identity, which had been erroneously proposed previously. This alternative mode of cyclization provides one with more flexibility for the chemoenzymatic synthesis of novel analogues of cyclic *ADP*-ribose, a potent mediator of calcium mobilization in many mammalian and invertebrate tissues.

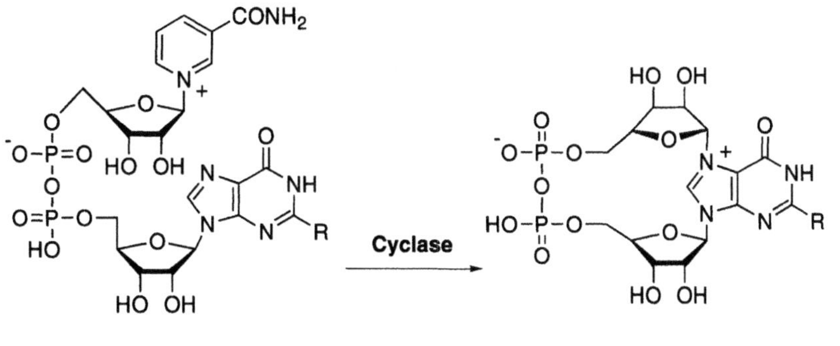

144: $R = NH_2$ **146:** $R = NH_2$

145: $R = H$ **147:** $R = H$

Scheme 38

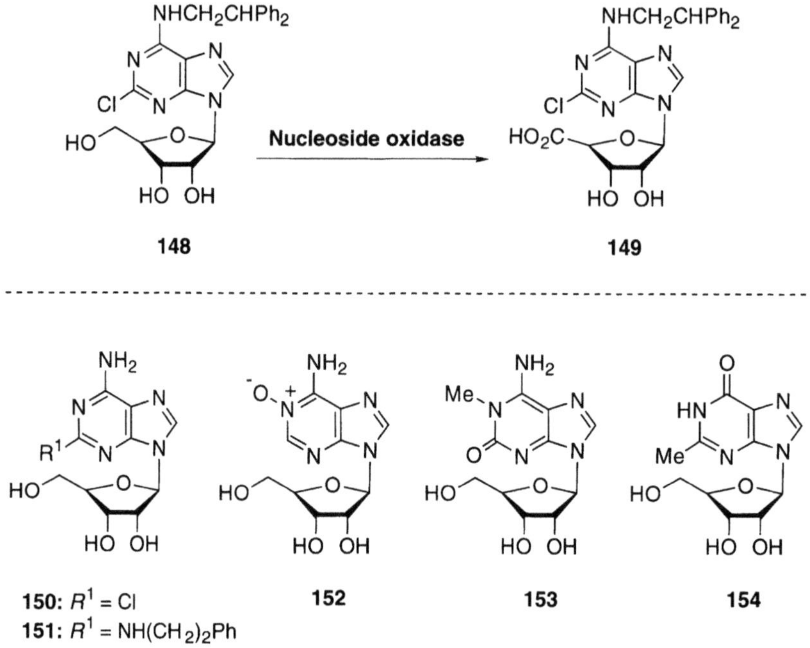

Scheme 39

Enzymatic oxidation of a nucleoside precursor offers potential for an improved synthesis (in comparison with a chemical one): it is environmentally clean and avoids the need for protection of other functional groups. An effective and practical enzymatic procedure [59] for the preparation of carboxylic nucleoside derivative **149** with nucleoside oxidase in high chemical yields from hydroxyl-unprotected nucleoside **148** has been reported (Scheme 39). Substrate specificity of this enzyme was also studied through a number of nucleoside 5′-carboxylates on a preparative scale. Nucleoside oxidase from *Stenotrophomonas maltophilia* (FERM BP-2252) has been developped for the analytical determination of nucleosides (assessing food freshness), but this is the first report for preparative synthesis of nucleoside 5′-carboxylates. It has been demonstrated that the enzyme is tolerant of different functionalities on the base, especially at the 2-position, including chloro (**150**) or phenylethyl amino (**151**) groups. Furthermore, the nitrogen at position 1 can be modified to the N-oxide (**152**) or methylated in the case of methyl isoguanine (**153**). Interestingly, in the inosine series, with 2-methylinosine (**154**) as a substrate, the reaction tended to stop at the 5′-aldehyde intermediate which was not further oxidized.

A new synthetic approach to nucleoside analogues based on enzyme-catalyzed aldol condensations has been reported [60]. This enzymatic route provides a new pathway to nucleosides with novel structures at the sugar and/or the base moiety and is outlined in Scheme 40. The racemic aldehyde component **156** used in the kinetically controlled enzymatic synthesis was obtained from glycidaldehyde diethyl acetal (**155**). The latter was treated with adenine to generate 3-adenyl-2-hydroxypropanal diethyl acetal, which was deprotected to form the free aldehyde

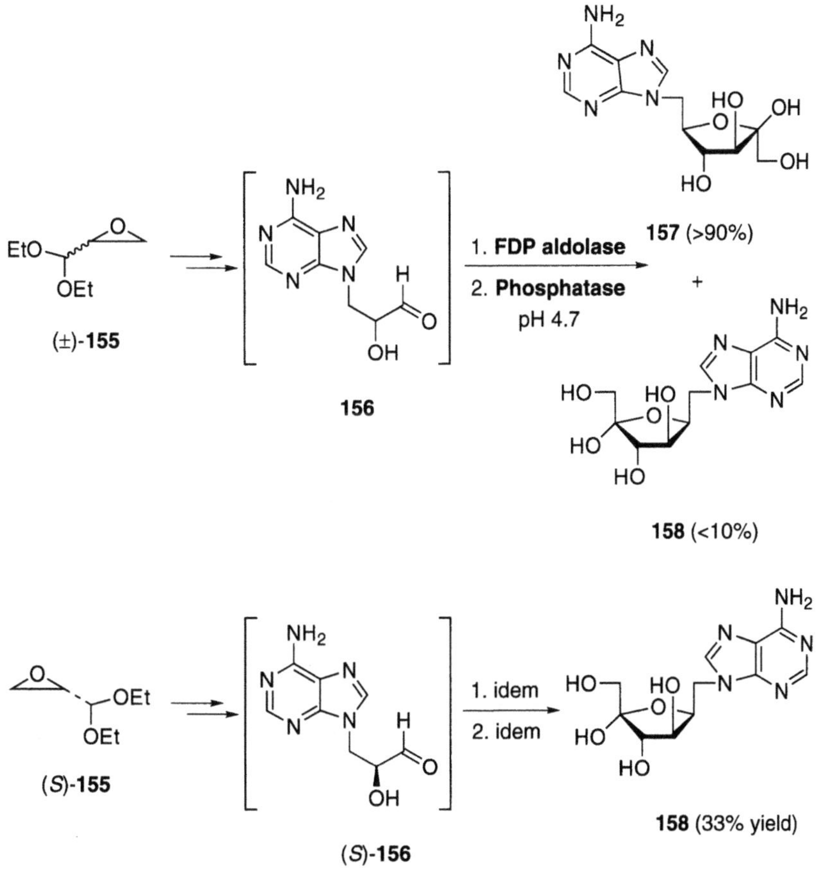

Scheme 40

156 *in situ*. Then FDP aldolase from rabbit muscle was added, thus initiating the aldolic enzymatic process. After the reaction was complete, the phosphate moiety was cleaved with acid phosphatase to afford compound **157** in 20% yield and a minor amount of product **158** ($\leq$10%). However, in a separate synthesis of **158**, enantiomerically pure (*S*)-**156** was prepared from (*S*)-**155** and used as a substrate for the enzymatic reaction.

Concluding Remarks

Although enzyme-catalyzed glycosyl transferring reactions have been the predominant type of biotransformations on nucleosides so far, the number of reports describing enzyme-mediated hydrolyses, protections, and resolutions has been increasing in the last two decades, especially since the introduction of organic solvent methodology. As the area continues to grow, new enzymatic procedures for the synthesis of many natural and unnatural nucleosides and intermediates will continue to be developed. Regioselective alkoxycarbonylation processes catalyzed by enzymes offers a new area of interest due to great flexibility and the possibility to provide practical routes to nucleoside carbonates of potential value. It is believed that more effective synthetic strategies based on combined chemical and enzymatic methods will have to be developed to tackle the new generation of problems associated with nucleosides, particularly those involved in modifications in both the heterocyclic base and the sugar moiety in order to avoid the drawbacks shown in certain applications such as agents effective against HIV and other viral infections or applications related with antisense oligonucleotide preparation. Moreover, biocatalysts are ecologically beneficial natural catalysts that offer the opportunity to carry out highly chemo-, stereo-, and regioselective synthesis of nucleosides which could not be performed by classical chemical methodologies.

References

[1] MacCoss M, Robins MJ (1990). In: Wilman DEV (ed) Chemistry of Antitumour Agents. Blackie and Son, UK, p 261; Robins RK, Kini GD (1990). In: Wilman DEV (ed) Chemistry of Antitumour Agents. Blackie and Son, UK, p 299; Robins RK, Revankar GR (1988). In: De Clercq E, Walker RT (eds) Antiviral Drug Development. Plenum, New York, p 11

[2] Sariaslani FS, Rosazza JP (1984) Enzyme Microb Technol **6**: 242; Whitesides GM, Wong C-H (1985) Angew Chem Int Ed Engl **24**: 617; Jones JB (1986) Tetrahedron **42**: 3351; Yamada H, Shimizu S (1988) Angew Chem Int Ed Engl **27**: 622; Csuk R, Glänzer BI (1991) Chem Rev **91**: 49; Santaniello E, Ferraboschi P, Grisenti P, Manzocchi A (1992) Chem Rev **92**: 1071; Turner NJ (1994) Nat Prod Rep **11**: 1; Azerad R (1995) Bull Soc Chim Fr **132**: 17; Theil F (1995) Chem Rev **95**: 2203; Schoffers E, Golebiowski A, Johnson CR (1996) Tetrahedron **52**: 3769; Schelhaas M, Waldmann H (1996) Angew Chem Int Ed Engl **35**: 2056; Roberts SM (1999) J Chem Soc Perkin Trans 1, 1; Gotor V (1999) Bioorg Med Chem **7**: 2189

[3] Azerad R, Sih CJ (1990) Enantioselective Synthesis using Biological Systems. In: Biocatalysis Special Issue, vol 3 (1–2); Meijer EM (1991) Biocatalysis in Organic Chemistry. In: Recl Trav Chim Pays-Bas Special Issue, vol 110 (5); Crout DHG, Roberts SM, Jones JB (1993) Enzymes in Organic Synthesis. In: Tetrahedron Asymm Special Issue, vol 4 (5–6); Kieslich K, Leuenberger HGW, Seebach D (1993) Biotransformations in Organic Chemistry: Principles and Applications. In: Chimia Special Issue, vol 47 (4) (Proc Int Bioorganic Symposium, Interlaken 1993); Crout DHG, Griengl H, Faber K, Roberts SM (1994) Proc European Symposium on Biocatalysis, Graz 1993. In: Biocatalysis Special Issue, vol 9 (1–4)

[4] Tramper J, VanderPlas HC, Linko P (1985) Biocatalysis in Organic Synthesis, Studies in Organic Synthesis, vol 22. Elsevier, Amsterdam; Porter R, Clark S (1985) Enzymes in Organic Synthesis, CIBA Foundation Symposium 111. Pitman, London; Schneider MP (1986) Enzymes as Catalyst in Organic Synthesis. In: NATO ASI Series C. Reidel, Dordrecht; Davies HG, Green RH, Kelly DR, Roberts SM (1989) Biotransformations in Preparative Organic Chemistry. The Use of Isolated Enzymes and Whole Cells Systems in Synthesis. Academic, London; Servi S (1992) Microbial Reagents in Organic Synthesis. In: NATO ASI Series C. Kluwer, Dordrecht; Wong C-H, Whitesides GW (1994) Enzymes in Synthetic Organic Chemistry. Elsevier, Oxford; Faber K (1997) Biotransformations in Organic Chemistry, 3rd ed. Springer, Berlin

[5] Drueckhammer DG, Hennen WJ, Pederson RL, Barbas III CF, Gautheron CM, Krach T, Wong C-H (1991) Synthesis 499; Gijsen HJM, Qiao L, Fitz W, Wong C-H (1996) Chem Rev **96**: 443

[6] Ferrero M, Gotor V (1999) Biocatalytic Synthesis of Steroids. In: Patel RN (ed) Stereoselective Biocatalysis. Dekker, New York, pp 579

[7] Prasad AK, Wengel J (1996) Nucleos Nucleot **15**: 1347

[8] Utagawa T (1999) J Mol Cat B: Enzymatic **6**: 215

[9] Prasad AK, Trikha S, Parmar VS (1999) Bioorg Chem **27**: 135

[10] Wolfenden R (1966) J Am Chem Soc **88**: 3157

[11] Nair V, Sells TB (1991) Synlett 753
[12] Dineva MA, Galunsky B, Kasche V, Petkov DD (1993) Bioorg Med Chem Lett **3**: 2781
[13] Kakefuda A, Shuto S, Nagahata T, Seki J-I, Sasaki T, Matsuda A (1994) Tetrahedron **50**: 10167; Seela F, Kaiser K (1988) Chem Pharm Bull **36**: 4153
[14] Robins MJ, Zon R, Hansske F, Wnuk SF (1997) Can J Chem **75**: 762
[15] Margolin AL, Borcherding DR, Wolf-Kugel D, Margolin N (1994) J Org Chem **59**: 7214
[16] Waldmann H, Heuser A, Reidel A (1994) Synlett 65
[17] Waldmann H, Reidel A (1997) Angew Chem Int Ed Engl **36**: 647
[18] Van Draanen NA, Koszalka GW (1994) Nucleos Nucleot **13**: 1679
[19] Zaitseva G, Zinchenko AI, Barai VN, Pavlova NI, Boreko EI, Mikhailopulo IA (1999) Nucleos Nucleot **18**: 687
[20] Sachdev HS, Starkovsky NA (1969) Tetrahedron Lett **9**: 733; Taunton-Rigby A (1973) J Org Chem **38**: 977
[21] Uemura A, Nozaki K, Yamashita J-I, Yasumoto M (1989) Tetrahedron Lett **30**: 3819
[22] Crout DHG, Dachs AM, Glover SE, Hutchinson DW (1990) Biocatalysis **4**: 177
[23] Baraldi PG, Bazzanini R, Manfredini S, Simoni D, Robins MJ (1993) Tetrahedron Lett **34**: 3177
[24] Baker DC, Haskell TH, Putt SR, Sloan BJ (1979) J Med Chem **22**: 273
[25] Singh HK, Cote GL, Sikorski RS (1993) Tetrahedron Lett **34**: 5201
[26] Damkjaer DL, Petersen M, Wengel J (1994) Nucleos Nucleot **13**: 1801
[27] Riva S, Chopineau J, Kieboom APG, Klibanov AM (1988) J Am Chem Soc **110**: 584
[28] Uemura A, Nozaki K, Yamashita J-I, Yasumoto M (1989) Tetrahedron Lett **30**: 3817
[29] Nozaki K, Uemura A, Yamashita J-I, Yasumoto M (1990) Tetrahedron Lett **31**: 7327
[30] Wong C-H, Chen S-T, Hennen WJ, Bibbs JA, Wang Y-F, Liu JL-C, Pantoliano MW, Whitlo M, Bryan PN (1990) J Am Chem Soc **112**: 945
[31] Zhong Z, Liu JL-C, Dinterman LM, Finkelman MAJ, Mueller WT, Rollence ML, Whitlo M, Wong C-H (1991) J Am Chem Soc **113**: 683
[32] Singh HK, Cote GL, Hadfield TM (1994) Tetrahedron Lett **35**: 1353
[33] de la Cruz A, Elguero J, Gotor V, Goya P, Martinez A, Morís F (1991) Synth Commun **21**: 1477
[34] de la Cruz A, Elguero J, Goya P, Martinez A, Gotor V, Morís F, de Clercq E (1992) J Chem Res (S) 216–217; J Chem Res (M) 1673
[35] Gotor V, Morís F (1992) Synthesis 626
[36] Morís F, Gotor V (1993) J Org Chem **58**: 653
[37] García-Alles LF, Gotor V (1998) Biotechnol Bioeng **59**: 163
[38] García-Alles LF, Gotor V (1998) Biotechnol Bioeng **59**: 684
[39] Ozaki S, Yamashita K, Konishi T, Maekawa T, Eshima M, Uemura A, Ling L (1995) Nucleos Nucleot **14**: 401
[40] Ciuffreda P, Casati S, Santaniello E (1999) Bioorg Med Chem Lett **9**: 1577
[41] Morís F, Gotor V (1992) J Org Chem **57**: 2490
[42] Morís F, Gotor V (1992) Tetrahedron **48**: 9869
[43] García-Alles LF, Morís F, Gotor V (1993) Tetrahedron Lett **34**: 6337
[44] García-Alles LF, Gotor V (1995) Tetrahedron **51**: 307
[45] Morís F, Gotor V (1993) Tetrahedron **49**: 10089
[46] Gotor V, Morís F, García-Alles, LF (1994) Biocatalysis **10**: 295
[47] García-Alles LF, Magdalena J, Gotor V (1996) J Org Chem **61**: 6980
[48] Magdalena J, Fernández S, Ferrero M, Gotor V (1998) J Org Chem **63**: 8873
[49] Magdalena J, Fernández S, Ferrero M, Gotor V (1999) Tetrahedron Lett **40**: 1787
[50] García-Alles LF, Gotor V (1999) J Mol Catal B: Enzymatic **6**: 407
[51] Binder WH, Kählig H, Schmid W (1995) Tetrahedron Asymm **6**: 1703
[52] Ohno M, Ito Y, Arita M, Shibata T, Adachi K, Sawai H (1984) Tetrahedron **40**: 145
[53] Wang J-Q, Archer C, Li J, Bott SG, Wang PG (1998) J Org Chem **63**: 4850
[54] Akita H, Chen CY, Kato K (1998) Tetrahedron **54**: 11011

[55] Morís F, Gotor V (1994) Tetrahedron **50**: 6927
[56] Waldmann H, Gabold S (1997) Chem Commun 1861
[57] Jungmann V, Waldmann H (1998) Tetrahedron Lett **39**: 1139
[58] Zhang F-J, Sih CJ (1995) Tetrahedron Lett **36**: 9289
[59] Mahmoudian M, Rudd BAM, Cox B, Drake CS, Hall RM, Stead P, Dawson MJ, Chandler M, Livermore DG, Turner NJ, Jenkins G (1998) Tetrahedron **54**: 8171
[60] Liu K-C, Wong C-H (1992) J Org Chem **57**: 4789

Received December 20, 1999. Accepted January 7, 2000

Lipase-Catalyzed Regio- and Stereoselective Acylation of Hydroxy Groups in Steroid Side Chains

Enzo Santaniello[1,*], **Patrizia Ferraboschi**[2], and **Shahrzad Reza-Elahi**[2]

[1] Dipartimento di Scienze Precliniche, LITA Vialba, Università degli Studi di Milano, I-20157 Milano, Italia

[2] Dipartimento di Chimica e Biochimica Medica, Università degli Studi di Milano, I-20133 Milano, Italia

Summary. Lipase from *Pseudomonas cepacia* (PCL) catalyzes the regio- and stereoselective acylation of primary and secondary hydroxy groups in steroid side chains under irreversible transesterification conditions with vinyl acetate in organic solvents.

Keywords. *Pseudomonas cepacia* lipase; Stereoselective acylation; Irreversible transesterification; Steroid side chain; Organic solvents.

Introduction

The lipase-catalyzed acylation of hydroxylated substrates in organic solvents is a well established method extensively applied to the synthesis of enantiomerically pure compounds [1–4]. For instance, under the conditions of irreversible transesterification in an organic solvent [5,6] we have successfully achieved the enantioselective resolution of a variety of 2-substituted alkanols [2]. In principle, the enzymatic method could be satisfactorily applied to substrates such as steroids that are highly insoluble in water. At the time we started our research on the regio- and stereoselective enzymatic acylation of functionalized steroid side chains, the only available examples of acylation on steroid alcohols referred almost exclusively to hydroxy groups located in the steroid rings [7–13].

Results and Discussion

We have reported for the first time that *Pseudomonas cepacia* lipase (PCL or lipase PS if the enzyme is from the Amano *Pseudomonas* species) catalyzes the stereoselective acylation of a hydroxy groups in a steroid side chain. The substrate was the (20*R*,*S*)-22-hydroxy steroid **1** which was acetylated with vinyl acetate in chloroform [14]. Starting from a 2:1 20*S*/20*R* ratio, in 68 h the epimeric ratio of the

* Corresponding author

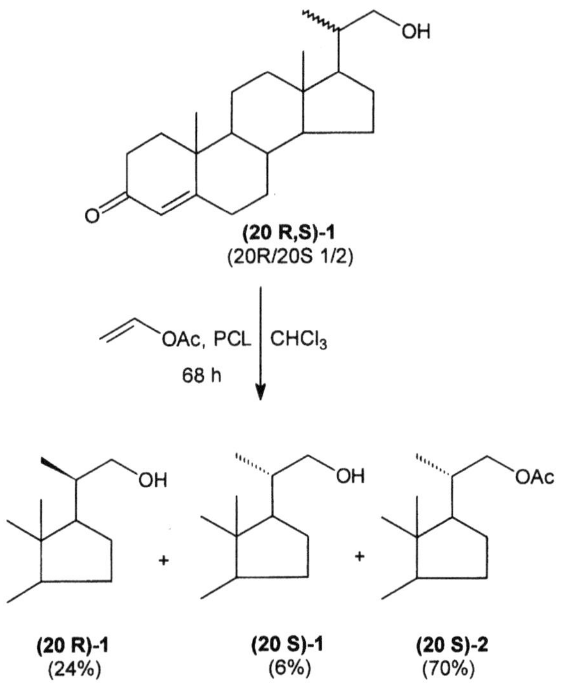

Scheme 1

unreacted alcohol **1** became 1:4. The preferential acetylation of the (20*S*)-isomer afforded the diastereomerically pure (20*S*)-acetate **2** (Scheme 1). For configurational analysis, a sample of epimerically pure (20*S*)-alcohol **1** was prepared from commercially available stigmasterol in a few steps [14].

The relatively slow reaction of the (20*R*)-alcohol and the stereochemical outcome of the enzymatic reaction may be, at least in part, determined by the steric hindrance of the 13-β methyl group. In fact, chemical transformations at the 22,23-position are influenced in the same way, as shown by the epimerization of the 22-aldehyde that, on reduction, determines the $20S/20R = 2{:}1$ ratio of the substrate **1**. Starting from the above considerations, we have examined a further primary alcohol function located in a steroid side chain at position 25 with the same stereogenic center as **1**. (25*R*,*S*)-26-Hydroxycholesterol (**3**) was subjected to the reaction with lipase and vinyl acetate in a chloroform/tetrahydrofuran mixture [15]. After 1 h, 30% of the (25*S*)-26-acetate **4** was formed, thus showing that the enzymatic reaction was also highly regioselective, since no trace of the 3-acetate was observed. Similarly, although slower, the (20*S*)-3β,22-dihydroxy derivative **5** was regioselectively converted in 30 h (30% yield) to the 22-acetate **6** (Scheme 2).

A 70% conversion to **4** was reached in 3 h, and the unreacted (25*R*)-**3** was isolated. The configuration of the enzymatic products **3** and **4** was assigned by comparison with published data [16], and the results were in agreement with the configurational outcome of the enzymatic reaction on other 2-methyl alkanols as substrates [17] within the model recently proposed for the interpretation of the enantiopreference of *Pseudomonas capacia* lipase toward primary alcohols [18]. Finally, comparing the hindered C-22 with the more accessible C-26 alcohol, the

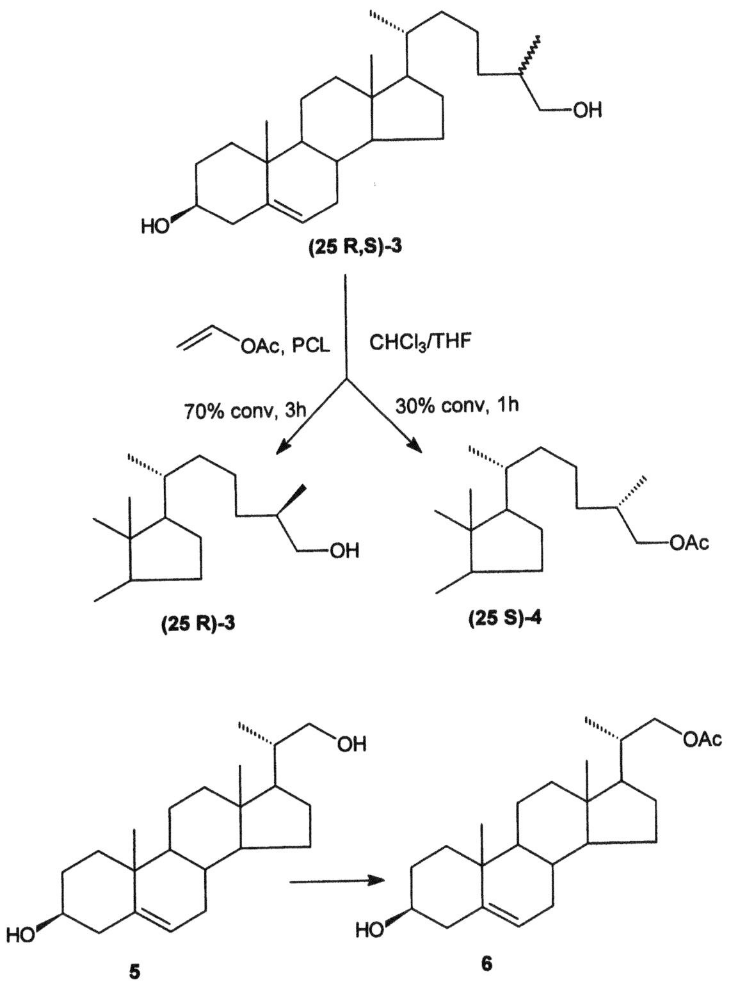

Scheme 2

enzymatic formation of the (25*S*)-acetate **4** is faster than the corresponding reaction on (20*R*,*S*)-22-hydroxy steroid **1**. This structural difference was confirmed by ^{1}H NMR analysis of the diastereomeric ratio of the two alcohols. In fact, in order to establish the 20*R*/20*S* ratio in **1** no derivatization was necessary, whereas the 26-alcohol **3** required the analysis of the corresponding *MTPA* ester [19].

The enzymatic acetylation of secondary hydroxy groups in a steroid side chain was also studied. In fact, 24-hydroxycholesterol, (24*R*,*S*)-**7** did not react under the usual conditions, and we assume that the isopropyl residue hinders the enzymatic acetylation. A similar negative result was obtained with the phenylsulfonyl alcohol **8** (Scheme 3).

For this reason, the 27-*nor*-(25*R*,*S*)-25-hydroxy-cholesterol **9** was prepared and subjected to the reaction with lipase and vinyl acetate in chloroform [20]. After 24 h, 30% of the 25-acetate **10** was formed, whereas a 70% conversion to **10** was reached in 48 h. The enzymatically prepared 25-alcohol **9** and the corresponding acetate **10** were obtained from two separate reactions in epimerically pure state as shown by the 500 MHz ^{1}H NMR spectra of their 25-MTPA esters. The configurations of the

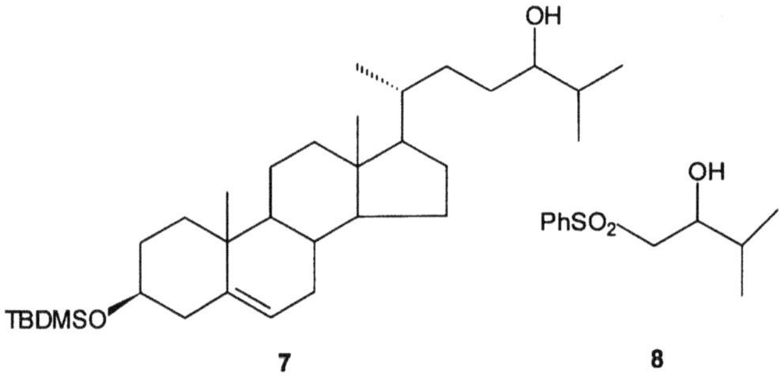

7 8

Scheme 3

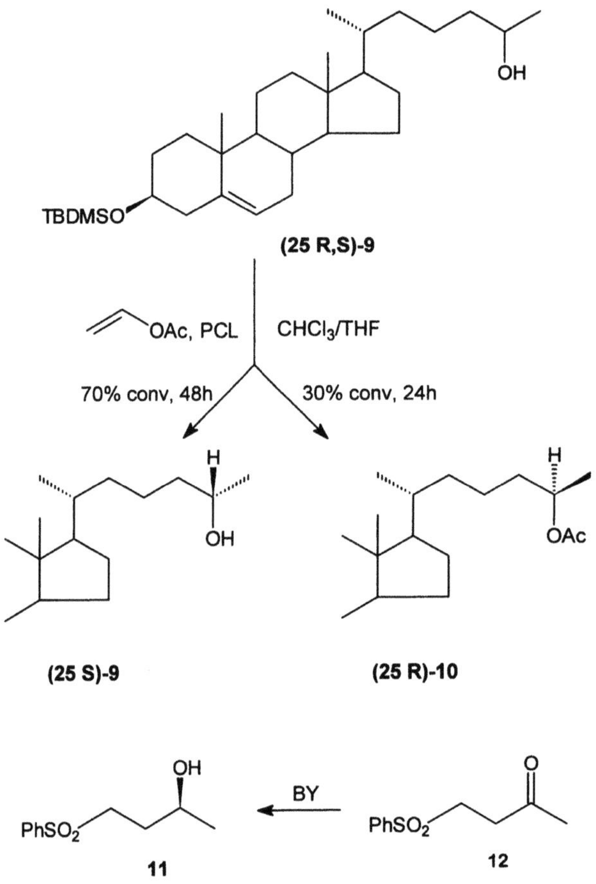

Scheme 4

enzymatic products were assigned by comparison with the resonances of a sample of a 25*S*-alcohol (**9**) synthesized from enantiomerically pure (*S*)-4-phenylsulfonyl-2-butanol (**11**), in turn prepared by baker's yeast-mediated reduction of 4-phenylsulfonyl-2-butanone (**12**) [20] (Scheme 4).

The results obtained from the steroidal alcohol **9** confirm that the 2-methyl-1-alkanols react faster than a secondary alcohol and that the stereochemical outcome of the enzymatic acylation of the alcohol **9** is in agreement with the configurational preference of the enzyme when a secondary alcohol is the substrate [21]. It should also be noted that the resolution of (25*R*,*S*)-26 acetate **4** and (25*R*,*S*)-25-acetate **10** by transesterification with methanol in *tert*-butyl methyl ether did not proceed satisfactorily. Finally, the results presented here show the first examples of enantioselective control of the lipase-catalyzed acylation of hydroxy groups present in steroid side chains and offer an additional chemoenzymatic access to the preparation of stereochemically pure sterogenic centers at these positions.

Experimental

Lipase-catalyzed transesterification; general procedure

A solution of 1 mmol of the steroid alcohol and 0.32 cm^3 vinyl acetate (3.46 mmol) in 5.5 cm^3 of a suitable organic solvent was added to 14 mg of solid lipase (31.5 U/mg) under stirring at room temperature for the time required for the desired conversion to the acetate. The reaction progress was monitored by TLC and GLC analysis (Hewlett Packard, model 5890/II, HP-5 capillary column, 280°C).

^1H NMR analysis of the stereochemical course of the enzymatic reactions

The ^1H NMR spectra (500 MHz) were registered using a Bruker AM 500 spectrometer. For the alcohol **1** no derivatization was required, and the 20*R*- and 20*S*-22-CH_2OH resonances were found at 3.58–3.63 and 3.65–3.72 ppm, respectively. The regioselective acylation of compound **5** was monitored following the formation of the 26-acetate **6** (multiplet centered at 3.87 ppm) and the unchanged multiplet centered at 3.50 ppm (3β-hydroxy group). The (*R*)-*MTPA* esters were required for establishing the epimeric ratio of the acetylation of 26-hydroxycholesterol **3**. The 25*R*- and 25*S*-26-CH_2OH resonances appeared as double doublets between 4.00–4.08 and 4.18–4.25 ppm and a doublet at 4.13 ppm, respectively. In the case of the 25-alcohol **9**, the significant resonances of the (*R*)-*MTPA* esters were those of the C-26 methyl group (1.23 and 1.30 ppm for 25*R* and *S*, respectively).

Acknowledgements

The work described here was financially supported by the *Ministero della Ricerca Scientifica e Tecnologica* (MURST PRIN *Biocatalisi e Bioconversioni*) and the *Consiglio Nazionale delle Ricerche* (CNR, *Target Project in Biotechnology*).

References

[1] Faber K, Riva S (1992) Synthesis 895
[2] Santaniello E, Ferraboschi P, Grisenti P (1993) Enzyme Microb Technol **15**: 367
[3] Fang J-M, Wong C-H (1994) Synlett 393
[4] Carrea G, Ottolina G, Riva S (1995) Trends Biotechnol **13**: 63
[5] Degueil-Castaing M, De Jeso B, Drouillard S, Maillard B (1987) Tetrahedron Lett **28**: 953
[6] Wang Y-F, Lalonde JJ, Momongan M, Bergbreiter DE, Wong C-H (1988) J Am Chem Soc **110**: 7200
[7] Riva S (1991) Enzymatic Modification of Steroids. In: Blanch HW, Clark DS (eds) Applied Biocatalysis, vol 1. Dekker, New York, p 179.
[8] Njar VCO, Caspi E (1987) Tetrahedron Lett **28**: 6549
[9] Riva S, Klibanov AM (1988) J Am Chem Soc **110**: 3291
[10] Adamczyk M, Chen Y-Y, Fishpaugh JR, Gebler JC (1993) Tetrahedron Asymm **4**: 1467

[11] Bertinotti A, Carrea G, Ottolina G, Riva S (1994) Tetrahedron **50**: 13165
[12] Baldessari A, Maier MS, Gros EG (1995) Tetrahedron Lett **36**: 4349
[13] Baldessari A, Bruttomesso AC, Gros EG (1996) Helv Chim Acta **79**: 999
[14] Ferraboschi P, Molatore A, Verza E, Santaniello E (1996) Tetrahedron Asymm **7**: 1551
[15] Ferraboschi P, Reza-Elahi S, Verza E, Santaniello E (1998) Tetrahedron Asymm **9**: 2193
[16] Uomori A, Seo S, Sato T, Yoshimura Y, Takeda K (1987) J Chem Soc Perkin 1, 1713
[17] Ferraboschi P, Casati S, De Grandi S, Grisenti P, Santaniello E (1994) Biocatalysis **10**: 279
[18] Weissfloch ANE, Kazlauskas RJ (1995) J Org Chem **60**: 6959
[19] Dale JA, Mosher HS (1973) J Am Chem Soc **95**: 512.
[20] Ferraboschi P, Pecora F, Reza-Elahi S, Santaniello E (1999) Tetrahedron Asymm **10**: 2497
[21] Kazlauskas RJ, Weissfloch ANE, Rappaport AT, Cuccia LA (1991) J Org Chem **56**: 2656

Received December 20, 1999. Accepted December 31, 1999

Comparative Study of Substrate- and Stereospecificity of Penicillin G Amidases from Different Sources and Hybrid Isoenzymes

Boris Galunsky, Karsten Lummer, and **Volker Kasche***

Arbeitsbereich Biotechnologie II, Technische Universität Hamburg-Harburg, D-21071 Hamburg, Germany

Summary. Four natural pencillin G amidase variants from different sources and two genetically constructed hybrid enzymes were produced and purified to homogeneity. The specificity constants of one enzyme (*E.* coli) were found to differ six orders of magnitude for hydrolytic transformations within a wide range of substrates. The substrate specificity of the homologous penicillin amidases was found to differ less than one order of magnitude for hydrolysis of the most specific and up to two orders of magnitude for the less specific substrates. The S_1'-substrate specificity in hydrolytic and transfer reactions (studied mainly with the *E. coli* enzyme) varied more than three orders of magnitude for the different substrates. The penicillin amidases were found to be *R*-specific in the S_1-binding site and *S*-specific in the S_1'-binding site. The S_1-stereoselectivity differs less than one order of magnitude for the different variants. The S_1'-stereoselectivity is more pronounced, increases with nucleophile specificity, and was found to differ up to three orders of magnitude in transfer reactions for the enzyme from *E. coli*. The observed variation of enatioselectivity for different penicillin amidases and one substrate can also be achieved by changes in temperature. Comparison of substrate- and stereospecificity of penicillin amidases from different sources and hybrid isoenzymes suggests that similar changes can be expected for enzyme variants derived by rational protein design or directed evolution.

Keywords. Homologous enzymes; Penicillin amidase (acylase); Stereospecificity; Substrate specificity.

Introduction

During the evolution, natural enzyme variants with a common function have been selected for favoured biological properties, e.g. high specificity (k_{cat}/K_m) and stereoselectivity for natural substrates at conditions in the main habitat of the microorganisms. Therefore, the enzyme properties differ for the homologous enzymes depending on their source. The evolution optimizes the biocatalysts with regard to their function *in vivo*, whereas their application as industrial biocatalysts requires tolerance to reaction conditions outside the physiological range and

* Corresponding author

transformation of substrates not encountered in nature [1]. During the last fifteen years much work has been done to optimize enzymes for industrial applications by rational protein design [2–5] or directed evolution [6–10]. Despite the progress in improving enzymes from a biotechnological point of view, fundamentally altered properties have hardly been achieved. The question on the relation between function and basic biocatalytic properties on one hand and structural parameters as the amino acid sequence on the other one still remains actual. Comparison of data on substrate- and stereoselectivity of homologous enzymes and mutation studies may provide insight into the role of selection in optimizing the kinetic behaviour of native enzymes and to what extent it could be modulated.

The family of pencillin G amidases (PA; EC 3.5.1.11) is suitable for such an analysis. These enzymes catalyze the hydrolysis of benzylpenicillin (**1**) to phenylacetic acid (**2**) and 6-aminopenicillanic acid (**3**) – one of the key intermediates for the production of semisynthetic β-lactam antibiotics [11]. PA can also catalyze the synthesis of these antibiotics [12, 13]. Due to the biotechnological interest, different natural and artificial variants of this enzyme have been produced [14–20]. The genes encoding penicillin G amidase in *E. coli*, *K. citrophila*, *P. rettgeri*, *A. faecalis*, and *A. viscosus* have been cloned and sequenced. The nucleotide sequences encode precursors that are processed to two nonidentical peptide chains (A and B) [11]. The homology in the primary structure of all compared enzymes decreases with increasing genetic distances between the different species which are supposed to be diverged from a common evolutionary ancestor [21]. Despite the large differences in the homology between the most distant variants (up to 70%), multiple sequence alignment data reveal in both A- and B-chains segments of strongly conserved amino acid sequences essential for the catalytic function of PA. The biological function of the penicillin amidase family is still unknown. The same applies for whether it is controlled by the same groups that control the biotechnical functions as a biocatalyst for benzylpenicillin hydrolysis and production of the penem nucleus (**3**) [22, 23]. The penicillin amidases have a wide range of substrate specificity and differ in properties important for existing and possible new industrial applications. The 3D-structure of the *E. coli* enzyme reveals a single amino acid (*Ser* B1) catalytic site [24], and supports its classification as an *Ntn*-hydrolase [25]. The 3D-structure confirms also the acyl-enzyme reaction mechanism [26, 27]. The same mechanism has been inferred for the other compared enzymes [11].

In this study we compare quantitative data on substrate- and stereoselectivity of different natural and genetically constructed hybrid enzyme variants. The results can yield information on how much these properties can be changed by directed or random mutagenesis.

Results and Discussion

The molar concentrations of the penicillin amidases were determined by active site titration [28]. Active site titration curves for all studied PA variants were almost identical which implies that only their N-terminal *Ser* B1 are accessible for acylation with phenylmethylsulfonyl fluoride (**4**).

Specificity constants in hydrolytic reactions

The S_1-subsite specificity[1] of penicilin G amidase is mainly restricted to phenylacetic acid and its derivatives [11], although other acyl moieties as acetyl, 2-benzoxazolon-3-yl-acetyl, or mandelyl are also accepted [30–33]. Most of the Pa-catalyzed transformations published so far have been done with the *E. coli* enzyme. For the substrates we studied with *E. coli* PA and presented in Fig. 1, the specificity constants for equilibrium controlled hydrolysis vary by 6 orders of magnitude. The observed differences between substrates varying in P_1 were predominantly due to larger differences in K_m. The S_1- and S_1'-subsite specificities of purified (homogeneous) natural and hybrid PA variants have been studied in detail for the hydrolysis of **1**, 6-nitro-3-phenylacetamido-benzoic acid (**11**) the less specific *R*-phenylglycine amide (**7**), *R*-phenylglycine methyl ester (**6**), *p*-hydroxy-*R*-phenylglycine amide (**9**), and 6-nitro-3-*R*-phenylglycylamidobenzoic acid (**10**). The results are shown in Fig. 2. The k_{cat}/K_m values for hydrolysis of **1** differ less than one order of magnitude for the different enzyme variants. The enzyme from *A. faecalis* has the highest specificity constant compared to the other natural enzymes. Similar data have been reported by other authors [34]. For the hydrolysis of **11**, **7**, and **6** the differences are larger – up to 2 orders of magnitude. Here differences in both the turnover number and the *Michaelis-Menten* constant account for the observed variation in the

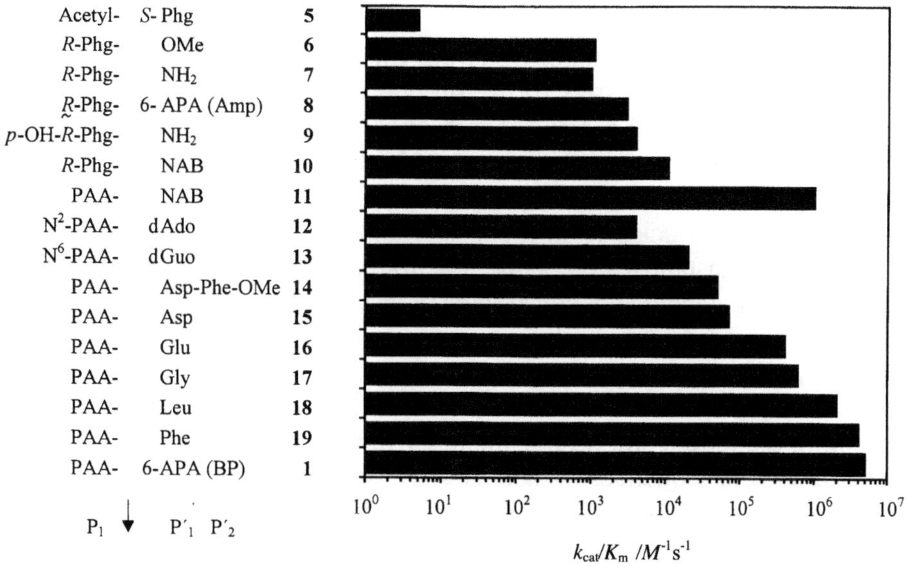

Fig. 1. Substrate specificity of affinity purified penicillin amidase from *E. coli* in equilibrium controlled hydrolytic reactions. Reaction conditions: phosphate buffer, $pH = 7.5$ ($I = 0.2$), 25°C. k_{cat} varies from 0.02 s^{-1} for **5** to 200 s^{-1} for **16**. K_m varies from 10 μ*M* for **1** to 30 m*M* for **6**. Abbreviations used: BP, benzylpenicillin; Amp, ampicillin; PAA, phenylacetic acid; Phg, phenylglycine; 6-APA, 6-aminopenicillanic acid; NAB, 6-nitro-3-aminobenzoic acid; dAdo, 2′-deoxyadenosine; dGuo, 2′-deoxyguanosine; three-letter symbols for the amino acids

[1] Using the convention for peptidases to designate the binding subsites [29]

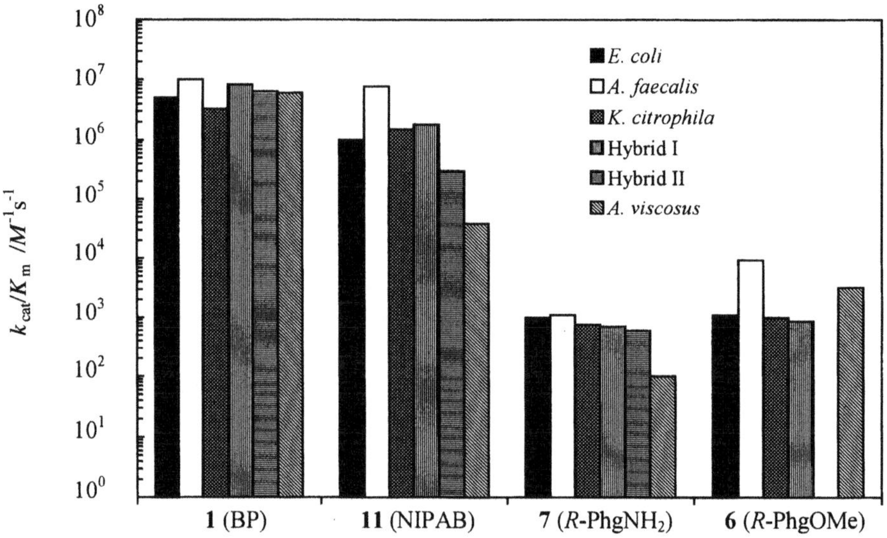

Fig. 2. Specificity constants of penicillin amidases from different sources and two genetically constructed hybrid enzymes (Hybrid I $A_{Kc}B_{Ec}$ and Hybrid II $A_{Ec}B_{Kc}$ for the hydrolysis of **1**, **11**, **7**, and **6**, at $pH = 7.5$ ($I = 0.2$) 25°C. All data were obtained with (affinity) purified enzymes. Both k_{cat} and K_m differ 3–4 fold for transformations of the same substrate catalyzed by different PA variants. Abbreviations: for **1**, **6**, and **7**, cf. Fig. 1; NIPAB: 6-nitro-3-phenylacetamido-benzoic acid

specificity constants of the homologous enzymes. The specificity constants differ also for the different processed forms of one enzyme. For *E. coli* PA, the last processed form with $pI = 6.6$ ($PA_{6.7}$) and an A-chain which is by 8 amino acids shorter than $PA_{7.0}$[2], has a specificity constant for hydrolysis of **1** lower by 30% [35]. The last processed form of *A. faecalis* PA with $pI = 5.3$ ($PA_{5.3}$), with the A-chain shortened for 3 amino acids from the C-end and converted N-terminal *Gly* a 30% larger specificity constant for the same reaction as compared to $PA_{5.5}$[3] [36].

We found that genetically constructed hybrids are active and have specificity constants for the studied hydrolytic reactions similar to those of the wild type enzymes [18]. The similarity of the kinetic behaviour of the homologous PA is not surprising. Besides the identity of *Ser* B1, *Gln* B23, *Ala* B69, *Asn* B241, and *Arg* B263 forming the tetrahedral arrangements of ligands of O_γ and the α-amino group [24], the amino acid sequences in their vicinity are strongly conserved as well: of 32 residues within 10 Å of *Ser* B1 only 3 differ in the different variants [17]. Changes at larger distances from the active site (the C-terminus of the A-chain) can also result in small activity changes, as discussed for the different processed forms. The smaller differences between the specificity constants of the homologous enzymes for hydrolysis of **1** compared to **11**, **7**, and **6** suggest that the transformation of one of the most specific substrates is probably a reaction congruent to the biological function of PA, which has been optimized during the evolution. Thus, there are hardly to be expected large differences in the catalytic behaviour of mutant enzymes for the most specific substrates. For catalyzing reactions of less specific substrates (more distant

---

[2] The dominating processed form with C-terminal Ala_{209} of the A-chain and $pI = 7.0$

[3] The processed form with C-terminal Ala_{202} of the A-chain and $pI = 5.5$

from the biological function), mutant enzymes with improved properties are more probable. The few site directed or randomly created PA mutants have shown lower specificity constants for hydrolysis of **1** [39–40].

S_1'-site specificity

The S_1'-specificity reflects the ability of penicillin G amidase to catalyze acyl transfer reactions [12]. It is broader and varies from penem and cephem nuclei to α- and β-amino acids, their phosphonic and phosphonous analogs, peptides, aromatic and aliphatic amines and alcohols, and nucleosides. These have been used (also on an industrial scale) for the kinetically controlled synthesis of amoxicillin, ampicillin, cephadroxil, and cephalexin, for deprotection in peptide and nucleotide synthesis, and for pro-drug activation in cancer therapy [13, 40–46]. We studied the nucleophile specificity in both hydrolytic and synthetic directions mainly with the *E. coli* enzyme. In Fig. 1, the specificity constants for hydrolytic transformation of substrates with phenylacetic acid in P_1-position and different moieties in positions P_1' and P_2' are compared. The k_{cat}/K_m values vary by more than 3 orders of magnitude for the different leaving groups. The few quantitative data on S_1'-specificity published by other authors reveal specificity constants of the same order of magnitude [41].

Interactions in the S_1'-binding subsite depend on electrostatic interactions between the basic amino acid residue in S_2'-position (most probably *Arg* A145) and the unprotected carboxyl groups of amino acid and β-lactam structures [12, 41, 47]. This has been observed also with *A. faecalis* PA [33]. In kinetically controlled synthetic reactions where the enzyme catalyzes the transfer of an acyl moiety to a nucleophile, the measure for the S_1'-specificity is the transferase to hydrolase ratio k_T/k_H [12]. We determined its valuewith *E. coli* PA for transfer of phenylacetyl or phenylglycyl moieties from phenylacetyl-*Gly* **17**, **7**, and **6** to amino acids or peptides. For the different substrates this ratios differ up to 3 orders of magnitude (1×10^1–3×10^4). For one nonspecific acyl moiety, k_T/k_H was found to have a value typical for a transferase (2×10^5), [48]. The *E. coli* enzyme was shown to be better for the acyl transfer to cephem nuclei than the PA from *A. faecalis.* Here comparative studies with other natural and hybrid penicillin G amidases are necessary to analyze (together with 3D-structures of enzyme-substrate complexes) a possible improvement of the transferase properties of the enzymes. This can be relevant for the synthesis of β-lactam antibiotics and for new biotransformations where C–N and C–O bonds are synthesized or hydrolyzed.

Stereoselectivity

The enantioselectivity of an enzyme E[4] should be >100 or <0.01 to obtain pure and *R*-enantiomers, respectively, at the end-point in biotransformations [49, 50]. PA stereospecificity is also very important for its biotechnical application. *E. coli*

[4] The quantitative measure for enzyme stereospecificity is the enantioselectivity or enantiomeric ratio *E*. It is defined as the ratio of the specificity constants for hydrolysis of enantiomeric *S*- and *R*-substrates or as the ratio of the transferase to hydrolase ratios for acyl transfer to enantiomeric *S*- and *R*-nucleophiles.

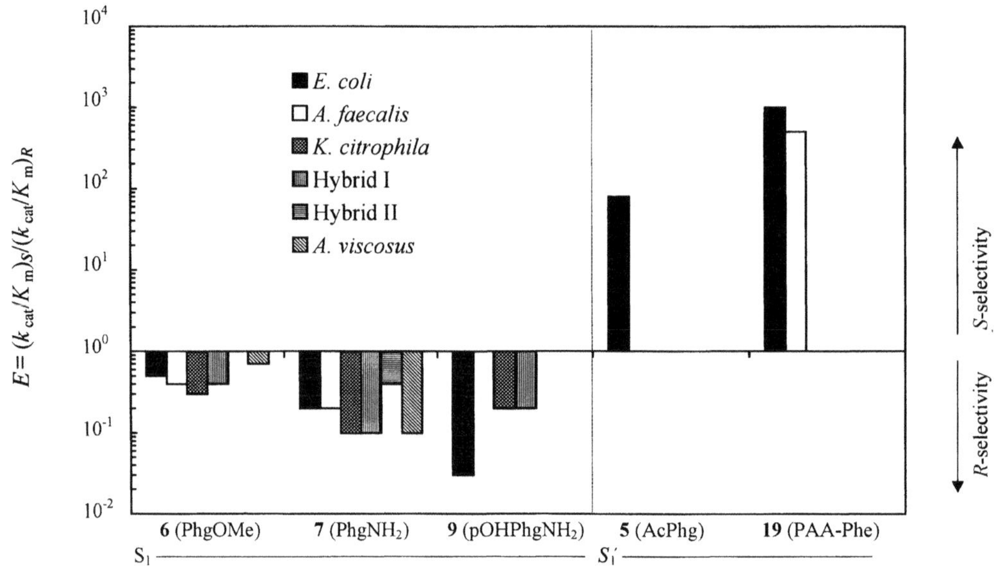

Fig. 3. S_1-subsite stereoselectivity of PA from different sources and two genetically constructed hybrid enzymes (Hybrid I $A_{Kc}B_{Ec}$ and Hybrid II $A_{Ec}B_{Kc}$), expressed as the enantiomeric ratio for various substrates hydrolysed at $pH = 7.5$ ($I = 0.2$) and 25°C. For comparison, the S_1'-subsite stereoselectivity is given for 2 substrates hydrolyzed at the same conditions by *E. coli* and *A. faecalis* enzymes. Abbreviations as in Fig. 1

PA has been used for the production of chiral synthons in peptide and nucleotide synthesis [41–44] and for kinetically controlled enantioselective acylation during loracarbef synthesis [51]. Considering the 3D-structure of *E. coli* PA and the sequence homology of the enzymes from different sources, very similar 3D-structures of all variants can be predicted [17]; thus, similar enantioselectivity could be expected.

S_1-enantioselectivity was studied with purified homologous penicillin G amidases only for a limited number of chiral substrates. The results are shown in Fig. 3. With the same enzyme, E was found to differ about one order of magnitude for the different substrates. A similar difference in enantioselectivity was observed for the transformation of one substrate by different PA variants. Differences mainly in k_{cat}, *i.e.* the activation energy, account for the observed changes.

S_1'-enantioselectivity was studied in both hydrolytic and transfer reactions, mainly with the *E. coli* enzyme. For comparison, in Fig. 3 the enantiomeric ratios for the hydrolytic transformations of two substrates catalyzed by two PA variants are given. S_1'-enantioselectivity is more pronounced and increases with nucleophile specificity. As for the S_1-stereoselectivity, differences mainly in k_{cat} account for differences in E. Literature data for the hydrolysis of phenylacetic acid derivatives of α-amino acids, peptides, β-amino acids, and alcohols reveal differences in E up to 3 orders of magnitude [41]. With the *E. coli* enzyme this was observed also in transfer reactions where $(k_T/k_H)_S/(k_T/k_H)_R$ differ up to 3 orders of magnitude (0.5–500) for the different substrates.

The penicillin G amidase was found to be R-specific in the S_1- and S- specific in the S_1'-binding subsites. This observation demonstrates that an enzyme consisting of homochiral amino acids can simultaneously catalyze reactions of two opposite chiral centres. This applies for all studied PA variants. Since the most specific substrates of PA [41] have no chiral moieties in P_1-position, it may imply that S_1-subsite stereospecificity is not important for the biological function. S'_1-subsite stereospecificity is more pronounced, moreover many of the most specific substrates [41] possess chiral centres in the P'_1-position. For directed modulation of E, however, more information on the structural determinants of enantiomer discrimination for different enzyme variants is necessary. Directed evolution studies have shown only moderate improvement of the enantiomeric ratio [52], whereas temperature variation have resulted in larger (more than order of magnitude) influence [30, 53]. Thus, alternative modulation of E by the reaction conditions should be considered as well.

Conclusions

1. The specificity constants of a homogeneous PA from *E. coli* were found to vary by 6 orders of magnitude for different $P_1 - P'_1 - P'_2$-substrates. The differences between substrates varying in P_1 is predominantly due to larger differences in K_m. This indicates that the catalytic power of this enzyme is influenced both by substrate binding and lowering of the activation energy of the transition state.
2. The enantioselectivity of the S_1- (*R*-specific) and S'_1-(*S*-specific) binding sites of this enzyme varied by up to 3 orders of magnitude for the hydrolysis of $P_1 - P'_1$ substrates. The differences are larger in the S'_1-subsite and in both cases mainly due to changes in k_{cat}.
3. Four natural and two hybrid variants of PA were purified to homogeneity. They differ up to 70% in primary structure. There is, however, a large homology of amino acid residues that are located near the B-chain N-terminal *Ser* in the *E. coli* enzyme, acylated in the reactions catalyzed by this enzyme.
4. The specificity constants of the PA variants for hydrolysis of one $P_1 - P'_1 - P'_2$ substrate or kinetically controlled synthesis with one nucleophile were found to differ by maximal two orders of magnitude. The changes decrease with the specificity constants for the substrates and are due to changes both in k_{cat} and K_m.
5. The enantioselectivity of the S_1- and S'_1-binding subsites of the enzyme variants varied up to one order of magnitude. Such a change for one PA variant and one substrate can also be achieved by changes in temperature.
6. Based on this we conclude that changes in substrate- and stereospecificity of new active PA variants created by directed or random mutations will be of the order of magnitude observed for the homologous enzymes, *i.e.* less than two orders of magnitude for specificity constants (the changes decreasing with the specificity constants) and less than one order of magnitude for stereoselectivities.

Experimental

Substrates

R- and *S*-enantiomers of **5**, **6**, and **7** were kindly provided by Röhm (Dermstadt, Germany); *R*- and *S*-enantiomers of **9** were kindly provided by DSM Research (Geleen, The Netherlands); *S*-enantiomers of **15** and **16** were kindly provided by Dr. *I. Stoineva*, Institute of of Organic Chemistry, Bulgarian

Academy of Sciences, Sofia, Bulgaria); **17**, the *S*-enantiomer of **18**, and *S*-and *R*-enantiomers of **19** were synthesized as described in Ref. [53]; *S,S*-diastereomer of **14** was synthesized as described in Ref. [44]; **12** and **13** were synthesized as described in Ref. [43]; **11** was synthesized as described in Ref. [54]; **1**, **4**, and **8** were purchased from Sigma (Desenhofen, Germany); all other chemicals were of analytical grade.

Preparation of PA variants

Penicillin amidase from *E. coli* and Hybrid I (A-chain from *K. citrophila* and B-chain from *E. coli*) were produced in recombinant *E. coli* strains and isolated by affinity chromatography using a biospecific adsorbent with an immobilized monoclonal antibody directed against the B-chain of *E-coli* enzyme as described in Refs. [18, 56]. The enzymers from *K. citrophila* (the gene in the plasmid pYKH5 was kindly provided by Dr. *Barbero*, Antibioticos, Leon, Spain), *A. feacalis* (the gene was amplified by the PCR-technique from the total *DNA* prepared from strain NCTC 415 by DSMZ, Braunschweig, Germany), and Hybrid II (A-chain from *E. coli* and B-chain from *K. citrophila*) were produced in recombinant *E. coli* strains as described in Refs. [30, 36, 18]. The Enzyme from *A. viscosus* was prepared using *DNA* from the wild strain ATCC 15294. These PA variants were purified from a concentrated homogenate by ion exchange chromatography using the same procedure as described previously [35, 36]. The purity and homogeneity of the prepared enzymes from different sources was analyzed by activity and protein stains of isoelectric focusing gels [35]. *E. coli* and *A. faecalis* enzymes used here were the proteolytically processed forms with isoelectric points of 7.0 and 5.3.

Active site titration

The molar concentrations of the PA variants were determined by active site titration with **4** [28] using **11** as a substrate.

Determination of k_{cat} and K_m.

The substrates were dissolved and diluted to the desired concentration in phosphate buffer (*pH* = 7.5, *I* = 0.2 *M*). The reaction was performed in this buffer at 25°C, initiated by addition of enzyme. The system was kept at constant temperature and *pH* value. The cleavage of **11** was followed spectrophotometrically at 380 nm by 6-nitro-3-aminobenzoic acid release [57]. With the other substrates, samples were withdrawn at different times and analysed by HPLC using a RP 8 column. The elution conditions and wavelengths used to quantitate the products are described in Refs. [58, 30, 44]. *Eadie-Hofstee* plots were used to determine k_{cat} and K_m.

Determination of k_T/k_H

The substrates (**6**, **7**, and **17**) and the nucleophiles (amino acids and peptides with free carboxyl function) were dissolved in carbonate buffer (*pH* = 9, *I* = 0.2 *M*). The ratio was determined in this buffer at 25°C. Samples were withdrawn and analysed by HPLC using a RP 18 column. Gradient elution was used as described in Ref. [30]. The peak areas at 225 nm were used to quantitate the products. k_T/k_H was calculated from the initial rates of condensation and hydrolysis product formation as described in Ref. [12].

Acknowledgements

This work has been supported by the *Deutsche Forschungsgemeinschaft* (Graduiertenkolleg Biotechnologie, GK 95-3-98).

References

[1] Rubingh DN (1997) Curr Opin Biotechnol **8**: 417
[2] Takagi H (1993) Int J Biochem **25**: 307
[3] Joyet P, Decklerk N, Gaillardin C (1992) Bio-Technology **10**: 1579
[4] You L, Arnold FH (1994) Protein Eng **9**: 77
[5] Rubingh DN (1996) Engineering Proteases with Improved Properties for Detergents. In: Savage L, Southborough MA (eds) Enzyme Technology for Industrial Applications. IBC, p 98
[6] Stemmer WPC (1994) Nature **370**: 389
[7] Stemmer WPC (1994) Proc Natl Acad Sci **91**: 10747
[8] Arnold FH (1996) Chem Eng Sci **51**: 5091
[9] Arnold FH, Moor JC (1997) Adv Biochem Eng **58**: 1
[10] Moore JC, Arnold FH (1996) Nat Biotechnol **14**: 458
[11] Virden R (1990) Biotechnol Gen Eng Rev **8**: 189
[12] Kasche V (1986) Enzyme Microb Technol **8**: 4
[13] Brugging A, Roos EC, de Vroom E (1998) Organic Process Research and Development **2**: 128
[14] Schumacher G, Sizmann D, Haug H, Buckel P, Bock A (1986) Nucleic Acids Res **14**: 5713
[15] Barbero JL, Buesa JM, Gonzales de Buitrago G, Mendez E, Perez Aranda A, Garcia JL (1986) Gene **49**: 69
[16] Ljubijankic C, Konstantinovic M, Glisin V (1992) DANN Seq **3**: 195
[17] Verhaert RMD, Riemens AM, van der Laan J-M, van Duin J, Quax WJ (1997) Appl Environ Microbiol **63**: 3412
[18] Piotraschke E, Nurk A, Galunsky B, Kasche V (1994) Biotechnol Lett **16**: 119
[19] Martin L, Prieto MA, Cortes E, Garcia JL (1995) FEMS Microbiol Lett **125**: 287
[20] Ohashi H, Katsuna Y, Hashizume T, Abe SN, Kajiura H, Hattori H, Kamei T, Yano M (1988) Appl Environ Microbiol **54**: 2603
[21] Olsen GJ, Woese CR and Overbeek R (1994) J Bacteriol **176**: 1
[22] Shewale JG, Sivaraman H (1989) Process Biochem **24**: 146
[23] Shewale JG, Deshpande BS, Sudhakaran VK, Ambedkar SS (1990) Process Biochem **25**: 97
[24] Duggleby H, Tolley SP, Hill CP, Dodson EJ, Dodson G, Moody PCE (1995) Nature **373**: 264
[25] Brannigan JA, Dodson G, Duggleby H, Moody PCE, Smith JL, Tomchik DR, Murzin AG (1995) Nature **378**: 416
[26] Konecny J (1981) Biotechnol Lett **3**: 112
[27] Haufler U (1987) PhD Thesis, University of Bremen, Germany
[28] Svedas VK, Margolin AL, Sherestyuk CF, Klesov AA, Berezin IV (1977) Bioorgan Khim (Russ) **3**: 546
[29] Schechter I, Berger A (1967) Biochem Biophys Res Commun **27**: 157
[30] Kasche V, Galunsky B, Nurk A, Piotraschke E, Rieks A (1996) Biotechnol Lett **18**: 455
[31] Niersbach H, Kühne A, Tischer W, Weber M, Wedekind F, Plapp R (1995) Appl Microbiol Biotechnol **43**: 679
[32] Stambolieva N, Mincheva Z, Galunsky B, Kalcheva V (1992) Enzyme Microb Technol **14**: 496
[33] Vojtisek V, Slezak J (1975) Folia Microbiol (Praha) **20**: 224
[34] Svedas, V, Guranda D, van Langen L, van Rantwijk F, Sheldon R (1997) FEBS Lett **417**: 414
[35] Kasche V, Haufler U, Markowsky D, Ziech A, Galunsky B (1987) Ann NY Acad Sci **501**: 97
[36] Ignatova Z, Stoeva S, Galunsky B, Hörnle C, Nurk A, Piotraschke E, Voelter W, Kasche V (1998) Biotechnol Lett **20**: 997
[37] Prieto I, Martin J, Arche R, Fernandez P, Perez-Aranda A, Barbero JL (1990) Appl Microbiol Biotechnol **33**: 553
[38] Williams JA, Zuzel TJ (1985) J Cell Biochem **9B**: 99
[39] Martin J, Prieto R, Barbero JL, Perez-Gil J, Mencheno JM, Arche R (1990) Biochim Biophys Acta **1037**: 133

[40] Forney LJ, Wong DCL, Ferber DM (1989) Appl Environ Microbiol **55**: 2550
[41] Waldmann H, Sebastian D (1994) Chem Rev **94**: 911
[42] Svedas VK, Savchenko MV, Beltser AI, Guranda DF (1996) Ann NY Acad Sci **799**: 659
[43] Waldmann H (1989) Tetrahedron Lett **30**: 3057
[44] Dineva M, Galunsky B, Kasche V, Petkov D (1993) Bioorg Med Chem Lett **3**: 2781
[45] Stoineva I, Galunsky B, Lozanov VS, Ivanov IP, Petkov DD (1992) Tetrahedron **48**: 1115
[46] Vrudhula VM, Senter PD, Fischer KJ, Wallace PM (1993) J Med Chem **36**: 919
[47] Alkema WBL, Prins AK, Janssen DB (1999) 4th International Symposium on Biocatalysis and Biotransformations, Giardini Naxos-Taormina, Italy, Abstract P112
[48] Stambolieva N, Mincheva Z, Galunsky B (1998) Biocat Biotrans **16**: 225
[49] Collins AN, Sheldrake GN, Crosby J (eds) (1992) Chyrality in Industry. Wiley, Chichester
[50] Rakels JLL, Paffen MT, Straathof AJJ, Heijnen JJ (1994) Enzyme Microb Technol **16**: 791
[51] Zmiewski MJ, Briggs, BS Tompson AR, Wright IG (1991) Tetrahedron Lett **32**: 1621
[52] Bornscheuer UT, Altenbuchner J, Meyer HH (1998) Biotechnol Bioeng **58**: 554
[53] Galunsky B, Ignatova S, Kasche V (1997) Biochim Biophys Acta **1343**: 130
[54] Suyama TJ (1965) Chem Abstr **63**: 7095
[55] Daumy GO, McColl AS, Apostolakos D (1982) J Bacteriol **152**: 104
[56] Kasche V, Gottschlich N, Lindberg A, Niebuhr-Redder C, Schmieding J (1994) J Chromatog **660**: 137
[57] Kutzbach C, Rauenbusch E (1974) Hoppe-Seyler's Z Physiol Chem **354**: 45
[58] Kasche V (1985) Biotechnol Lett **7**: 877

Received December 20, 1999. Accepted (revised) February 4, 2000

Lipase-Catalyzed Resolution of Ibuprofen

Erik Henke, Sascha Schuster, Hong Yang, and **Uwe T. Bornscheuer***

Institut für Chemie und Biochemie, Abt. Technische Chemie und Biotechnologie, Universität Greifswald, D-17487 Greifswald, Germany

Summary. The resolution of ibuprofen by transesterification of its corresponding vinylester using lipase B from *Candida antarctica* is described. Compared to transesterification or hydrolysis of the ibuprofen ethyl ester ($E < 2$, 28–48 h), the reaction with vinylesters occurred significantly faster (1.5–5 h) and with considerably higher enantioselectivity ($E = 8$–39).

Keywords. Enantioselectivity; *Candida antarctica* B; Ibuprofen; Lipase; Organic solvent; Vinylester.

Introduction

2-Arylpropionic acids (profens) are known as a major group of nonsteroidal anti-inflammatory drugs (NSAID) used in the treatment of arthritis and related diseases [1, 2]. *In vitro* tests have shown that the anti-prostaglandin synthetase activity of profens resides almost exclusively in the (+)-(*S*)-enantiomers, yet all profens except naproxen are marketed as racemates [3]. In the case of ibuprofen (2-(4-isobutylphenyl)-propionic acid) it has been shown that the (*S*)-enantiomer is about 160 times biologically more active than its (*R*)-counterpart [4].

Lipases (EC 3.1.1.3) are very suitable enzymes for organic syntheses because they accept a wide range of non-natural substrates, are stable and active in organic solvents, do not require cofactors, and are readily available from several (micro)organisms in sufficient quantities. So far, the resolution of >1000 substances has been described, and it turned out that enantioselectivity toward secondary alcohols is usually very high. In contrast, primary alcohols as well as carboxylic acids are often difficult to resolve [5].

Several groups already have investigated the kinetic resolution of NSAID-precursors, but as outlined above, the enantioselectivity E [6] was only moderate in most cases due to the low enantioselectivity observed in the lipase-catalyzed hydrolysis of racemic carboxylic acid derivatives (*e.g.* a methyl ester) of NSAIDs. In contrary, esterification of the racemic carboxylic acids with simple alcohols (*e.g.* methanol) is hampered by an unfavourable equilibrium leading to long reaction times and low conversions. For instance, esterification of (*R,S*)-ibuprofen with *n*-propanol using lipase from *Candida rugosa* proceeded with $E = 3$ in isooctane.

* Corresponding author

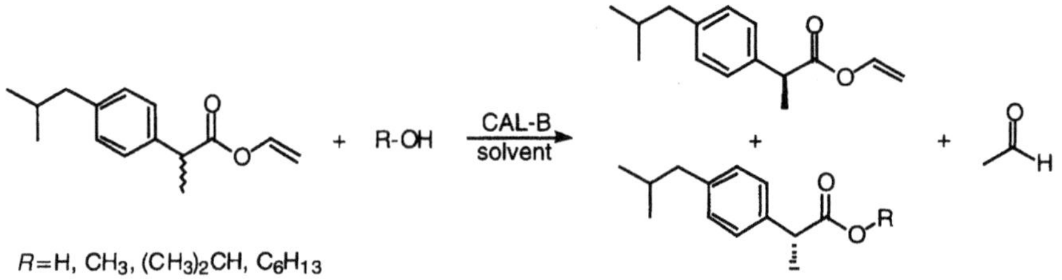

Scheme 1

In the presence of microemulsion, enantioselectivity was much higher ($E = 150$), but 10 days reaction time were necessary to achieve ~30% conversion [7]. A similar reaction in supercritical carbon dioxide as solvent gave an E-value of *ca.* 6 at <20% conversion [8]. In another example, two subsequent kinetic resolutions using lipase B from *Candida antarctica* (CAL-B) have led to the isolation of (*S*)-ibuprofen with 85% *ee* after the first resolution and 97.5% *ee* after the second step, which corresponds to an E-value of 17–20. The overall yield was 15% at a total reaction time of 72 h [9].

Recently, we have shown that the resolution of carboxylic acids can be efficiently performed by employing the corresponding vinyl esters, which are transesterified with a suitable alcohol by the aid of lipase B from *Candida antarctica* (CAL-B) [10]. This approach not only lead to significantly enhanced reaction rates as compared to the ethyl esters, but also enantioselectivity increased dramatically from *e.g.* $E = 6.5$ to $E > 100$ in the resolution of 2-phenylbutyric acid derivatives at reasonable reaction times. Moreover, this method also enabled an efficient synthesis of optically pure diastereomeric esters [11]. In both cases, high reaction rates stem from the fact that the vinyl alcohol generated during the transesterification tautomerizes to acetaldehyde, thus making the reaction irreversible. Preliminary computer-aided molecular modeling based on the structure of CAL-B suggests that the increase in enantioselectivity is due to a favorable interaction of the vinyl ester compared to ethyl ester in the binding pocket of CAL-B (data not shown). In this paper, we describe the application of this method for the resolution of ibuprofen (Scheme 1).

Results and Discussion

Initially, we have investigated the ability of three commonly-used lipases (from *Candida antarctica* (CAL-B), *Pseudomonas cepacia* (PCL), and *Candida rugosa* (CRL)) to resolve racemic ethyl or vinyl esters of ibuprofen either in a hydrolytic reaction or by transesterification in toluene as solvent.

For a fast and accurate determination of activity and enantioselectivity and the rapid evaluation of enzymes and reaction conditions, we used a reaction setup with 2 cm^3 reaction vials thermostated in a thermoshaker, which enabled us to use small amounts of substrates (100 μmol/reaction). Samples were analyzed by gas chromatography on a column coated with a chiral carrier, allowing the determination of enantiomeric excess of substrate and product as well as conversion in a single run.

PCL showed no activity towards (*R,S*)-ibuprofen vinyl ester (**1**) or (*R,S*)-ibuprofen ethyl ester (**2**), whereas CRL was only modestly active and enantiose-

Table 1. Resolution of ibuprofen vinyl ester (**1**) or ibuprofen ethyl ester (**2**) using CAL-B

Substrate	Nucleophile[a]	Reaction time (h)	Conversion (%)	Enantiomeric (%ee_S)[b]	excess (%ee_P)	E[c]
1	Methanol	2	52	68 (75)	69	12
1	2-Propanol	5	50	65 (63)	63	8
1	*n*-Hexanol	1.5	57	99 (99)	75	39
2	Methanol	28	33	8 (10)	21	<2
1	Water	11.5	86	99 (99)	16	12
2	Water	48	49	15 (16)	17	<2

[a] For reaction conditions with different nucleophiles see experimental section; [b](*S*)-configuration, values calculated from ee_p and conversion are given in parentheses; [c]calculated according to Ref. [6] using the program available at *http://www-orgc.tu-graz.ac.at*

lective ($E \sim 5$) in hydrolysis and also did not catalyze transesterifications, presumably because a non-immobilized preparation was used. In accordance with our previous results [10], CAL-B showed activity in both reactions (Table 1, Scheme 1). However, it is obvious from Table 1 that the reaction rate as well as the enantioselectivity differed considerably. The highest enantioselectivity ($E = 39$) and shortest reaction time (1.5 h) were observed for the transesterification of vinyl ester **1** with *n*-hexanol. The use of other alcohols as nucleophiles still proceeded much faster than hydrolysis, but *E* dropped to values between 8 and 12. In contrast, resolution of the ethyl ester gave almost racemic products at *E*-values below 2 independent whether transesterification with methanol or hydrolysis were performed. Thus, the only method to obtain highly optically pure ibuprofens is by using vinyl esters. This is also exemplified by the time course for the resolution using vinyl ester **1** (Fig. 1) or ethyl ester **2** (Fig. 2). In both reactions, the remaining ester had (*S*)-configuration.

Because the higher enantioselectivity was also observed in the hydrolysis of the vinyl ester, the vinyl group must be responsible for a more stereoselective binding

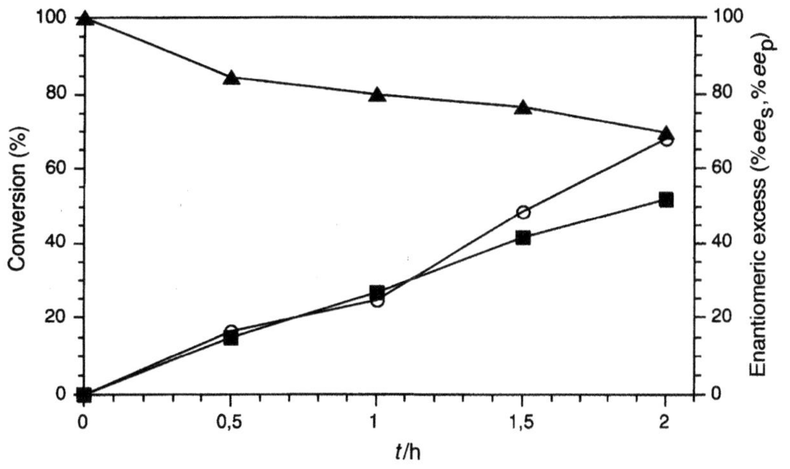

Fig. 1. Transesterification of ibuprofen vinyl ester (**1**) with methanol; ■: conversion, ○: enantiomeric excess of vinyl ester (ee_S), ▲: enantiomeric excess methyl ester (ee_P)

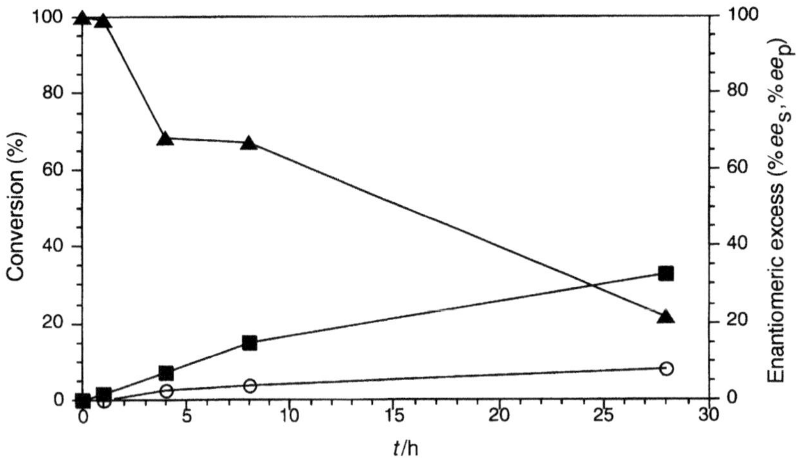

Fig. 2. Transesterification of ibuprofen ethyl ester (**2**) with methanol; ■: conversion, ○ enantiomeric excess ethylester (ee_S); ▲: enantiomeric excess methyl ester (ee_P)

of the substrate within the active site of CAL-B compared to the ethyl group. Although both molecules differ only slightly in terms of size, the vinyl group might lead to a better binding due to the presence of the double bond, *i.e.* by forming an additional hydrogen bond, thus causing a more favoured conformation of the vinyl ester. In addition, E increased with increasing chain length of the alcohol used, suggesting that the binding of the nucleophile also contributes to enantioselectivity.

Experimental

^{1}H NMR spectra were recorded at 250.1 and 500.1 MHz, ^{13}C NMR spectra at 62.9 and 125.7 MHz using Bruker NMR spectrometer in $CDCl_3$ with tetramethylsilane as internal standard. Gas chromatographic analyses were conducted on a Finnigan gas chromatograph equipped with a flame ionization detector using a *heptakis*-(2,3-di-O-methyl-6-O-pentyl)-β-cyclodextrin column (25 m × 0.25 mm, Prof. *W. A. König*, University of Hamburg, Germany). Immobilized lipase B from *Candida antarctica* (Chirazyme L-2, c.-f., C2, lyo.; 5000 U/g) and lipase from *Pseudomonas cepacia* (Chirazyme L-1, c.-f., lyo.; 5000 U/g) were donated by Roche Diagnostics, Penzberg, Germany. Lipase from *Candida rugosa* (Amano AY) was donated by Amano Inc., Nagoya, Japan. All chemicals were purchased from Fluka-Sigma-Aldrich, Deisenhofen, Germany. The results of elemental analyses agreed favourably with the calculated values.

(R,S)-Ibuprofen vinyl ester (**1**, $C_{15}H_{20}O_2$)

In 90 cm^3 vinylacetate (1.05 mol), 2 g ibuprofen (9.7 mmol), 348 mg palladium(II)acetate (1.55 mmol), and 54 mg potassium hydroxide (0.97 mmol) were dissolved. After refluxing the mixture for 15 h, the solids were filtered off and the residue was washed with 20 cm^3 vinyl acetate. The solvent was removed under vacuum before the remaining crude product was purified by flash column chromatography (silica gel, petrolether:diethylether = 20:1).

Yield: 2.10 g ibuprofen vinyl ester (7.48 mmol, 80%); ^{1}H NMR (500.1 MHz, $CDCl_3$), $\delta = 0.89$ (6H, d, $J = 6.5$), 1.52 (3H, d, $J = 7.1$), 1.81–1.87 (1H, m), 2.44 (2H, d, $J = 7.2$), 3.75 (1H, q, $J = 7.1$), 4.54 (1H, dd, $J = 1.6$, $J = 6.3$), 4.86 (1H, dd, $J = 1.6$, $J = 14.6$), 7.09–7.24 (4H, m) ppm; ^{13}C NMR (125.8 MHz, $CDCl_3$): $\delta = 18.50$, 22.46, 30.24, 44.96, 45.11, 97.92, 127.27, 129.51, 136.97, 140.89, 141.50, 171.90 ppm.

(R,S)-Ibuprofen ethyl ester (**2**; $C_{15}H_{18}O_2$)

In a $100\,cm^3$ round bottomed flask, 1 g (*R,S*)-ibuprofen, (4.88 mmol) were dissolved in $30\,cm^3$ ethanol. After addition of $50\,mm^3$ H_2SO_4 (0.98 mmol) the mixture was refluxed on an oil bath for 22 h. Ethanol was removed by distillation using a 20 cm vigreux column. The residual volume of approximately $10\,cm^3$ was dissolved in $50\,cm^3$ diethyl ether and washed with $100\,cm^3$ saturated sodium hydrogen carbonate solution. The aqueous layer was extracted twice with $30\,cm^3$ diethyl ether. The organic layers were combined, washed with $50\,cm^3$ water, and dried over anhydrous sodium sulfate. Solvent removal under vacuum yielded 1 g of colorless product (4.28 mmol, 87%) which needed no further purification (purity >99% as determined by GC and NMR analysis).

^{1}H NMR (500.1 MHz, $CDCl_3$): $\delta = 0.89$ (6H, d, $J = 6.5$), 1.20 (3H, d, $J = 7.2$), 1.48 (3H, d, $J = 7.1$), 1.86 (1H, m), 2.44 (2H, d, $J = 7.2$), 3.67 (1H, q, $J = 7.2$), 4.06–4.17 (2H, m), 7.09 (2H, d, $J = 8.1$), 7.15 (2H, d, $J = 8.1$) ppm; ^{13}C NMR (125.8 MHz, $CDCl_3$): $\delta = 14.53$, 19.02, 22.80, 30.59, 45.44, 45.56, 61.05, 127.52, 129.69, 138.28, 140.85, 175.19 ppm.

General procedure for the enzymatic transesterification of ibuprofen vinyl ester (**1**) *and ibuprofen ethyl ester* (**2**)

To $1\,cm^3$ of a solution of 100 m*M* **1** and 100 m*M* alcohol (500 m*M* alcohol in case of methanol and **2**) in dry toluene, 10 mg of the lipase preparation were added in a $2\,cm^3$ reaction vial and incubated at 40°C in a thermoshaker. Samples of $10\,mm^3$ were withdrawn from the reaction mixture, diluted with $1\,cm^3$ of toluene, and the lipase was removed by centrifugation. The diluted sample was analyzed directly by gas chromatography on a chiral column.

General procedure for the enzymatic hydrolysis of **1** *and* **2**

To $200\,mm^3$ of a solution of 500 m*M* **1** or **2** in toluene, 10 mg of the lipase preparation and $800\,mm^3$ sodium phosphate buffer (50 m*M*, *pH* = 7.5) were added in a $2\,cm^3$ reaction vial and incubated at 40°C in a thermoshaker. Samples of $5\,mm^3$ were withdrawn from the organic layer, diluted in $1\,cm^3$ of toluene, and the lipase was removed by centrifugation. The diluted sample was analyzed directly by gas chromatography on a chiral column.

Lipase-catalyzed synthesis of ibuprofen methyl ester

Racemic ibuprofen vinyl ester (**1**, 209 mg, 0.90 mmol) and $300\,mm^3$ methanol (240 mg, 7.5 mmol) were dissolved in $8\,cm^3$ toluene, and the mixture was heated to 40°C. The reaction was started by addition of 416 mg lipase CAL-B. After 5 h the immobilized lipase was filtered off, and the solvent was removed under vacuum. An aliquot of the residue was used to determine the conversion (40.5%) by gas chromatography. The remaining mixture of product and non-converted substrate was separated by flash column chromatography (silica gel, petrolether:diethyl ether = 35:1).

Yield: 64 mg ibuprofen methyl ester (0.29 mmol, 32%); $ee_p = 76\%$ *ee* (GC-analysis); $[\alpha]^{25}_{589nm} = -50.2°$ (corresponding to 77% ee_P; $CDCl_3$; $c = 6.800$; Ref. [13]: $[\alpha]^{25}_{589nm} = 64.6°$ ($CDCl_3$) for (*S*)-Ibuprofen methyl ester); 98 mg ibuprofen vinyl ester (0.42 mmol, 47%) $ee_S = 52\%$ *ee* (calculated from ee_p and conversion); $[\alpha]^{25}_{589nm} = +28.6°$ ($CDCl_3$, $c = 4.900$); $E = 12$.

Ibuprofen methyl ester: ^{1}H NMR (250.1 MHz, $CDCl_3$): $\delta = 0.82$ (6H, d, J = 6.6 Hz), 1.41 Hz (3H, d, $J = 7.2$ Hz), 1.77 (1H, septet, $J = 6.8$), 2.37 (2H, d, $J = 7.2$ Hz), 3.58 (3H, s), 3.70 (1H, q, $J = 7.2$ Hz), 7.00–7.18 (4H, m) ppm; ^{13}C NMR (62.9 MHz, $CDCl_3$), $\delta = 18.70$, 22.47, 30.25, 45.10, 45.11, 52.05, 127.20, 129.43, 137.82, 140.63, 175.30 ppm. NMR data of isolated vinyl ester matched with those of the racemic compound **1** as given above.

Acknowledgements

The authors are grateful for a grant by the German Research Foundation (DFG, Bonn, Germany), financial support by the *Fonds der Chemischen Industrie* (Frankfurt, Germany), and a postdoctoral stipend for Dr. *Hong Yang* from the *Alexander v. Humboldt Foundation* (Bonn, Germany). We also thank Prof. *W. A. König* (University of Hamburg, Germany) for providing the column for chiral analysis by gas chromatography.

References

[1] Adams SS (1992) J Clin Pharmacol **32**: 317
[2] Davies EF, Avery GS (1971) Drugs **2**: 416
[3] Mayer JM (1990) Acta Pharm Nord **2**: 197
[4] Adams SS, Bresloff P, Mason GC (1976) J Pharm Pharmacol **28**: 256
[5] Bornscheuer UT, Kazlauskas RJ (1999) Hydrolases in Organic Synthesis – Regio- and Stereoselective Biotransformations. Wiley-VCH, Weinheim
[6] Chen CS, Fujimoto Y, Girdaukas G, Sih CJ (1982) J Am Chem Soc **104**: 7294
[7] Hedström G, Backlund M, Slotte JP (1993) Biotechnol Bioeng **42**: 618
[8] Rantakylä M, Aaltonen O (1994) Biotechnol Lett **16**: 825
[9] Trani M, Ducret A, Pepin P, Lortie R (1995) Biotechnol Lett **17**: 1095
[10] Yang H, Henke E, Bornscheuer UT (1999) J Org Chem **64**: 1709
[11] Yang H, Henke E, Bornscheuer UT (1999) Tetrahedron Asymmetry **10**: 957
[12] Lobell M, Schneider MP (1994) Synthesis 375
[13] Ghislandi V, Manna A, Azzolina O, Gazzaniga A, Vercesi D (1982) Farmaco **37**: 81

Received December 20, 1999. Accepted January 12, 2000

Novel Carboxyl Esterase Preparations for the Resolution of Linalyl Acetate

Mateja Pogorevc[1], **Ulrike T. Strauss**[1], **Marianne Hayn**[2], and **Kurt Faber**[1,*]

[1] Institute of Organic Chemistry, University of Graz, A-8010 Graz, Austria
[2] Institute of Biochemistry, University of Graz, A-8010 Graz, Austria

Summary. Biocatalytic resolution of the tertiary terpene alcohol (±)-linalool was accomplished *via* hydrolysis of its corresponding acetate ester using two highly enantiospecific enzymes ($E > 100$). The latter were identified in a crude cell-free extract of *Rhodococcus ruber* DSM 43338 and could be separated by (partial) protein purification. Since they showed opposite enantiopreference, they were termed (*R*)- and (*S*)-linalyl acetate hydrolase (LAH). The activity and selectivity of the enzyme preparations was markedly dependent on the fermentation conditions.

Keywords. Linalool; Enzymatic resolution; *tert*-Alcohol; Linalyl acetate hydrolase.

Introduction

Linalool is an important chiral terpene alcohol used in flavor and fragrance industry. Since its enantiomers differ in odor [1], their availability in optically pure form is desirable for flavor and fragrance composition. The (*R*)-(−)-enantiomer – also denoted as licareol – is a major constituent of *Cinnamonium camphora* and *Cayenne linaloe* extract, whereas the (*S*)-(+)-counterpart primarily occurs in coriander oil (coriandrol).

Several attempts to resolve (±)-linalool (**1a**) by means of biocatalytic methods have been reported to date. For instance, (±)-**1a** was shown to be a non-substrate for several microbial lipases in esterification reactions [2, 3]. In an analogous fashion, a broad range of esterases, proteases, and lipases of various origins such as microbial, plant, and mammalian sources were found to be inactive in the hydrolysis of the corresponding acetate (±)-**1** [4]. This is not surprising, since tertiary alcohols and esters thereof are generally not accepted as substrates by the most commonly employed carboxyl ester hydrolases [5, 6]. As a consequence, several studies in search for novel microbial enzymes being capable to hydrolyze linalyl acetate in a selective fashion have been initiated. When using whole cells of *Aspergillus niger*, the hydrolysis succeeded without noticeable enantioselectivity ($E < 2$) [7]. In addition, several by-products arising from hydroxylation and rearrangement of the carbon skeleton were formed in significant amounts. Similar activities have been reported using whole resting cells of *Bacillus subtilis* [8]. Although (*S*)-linalool was

* Corresponding author

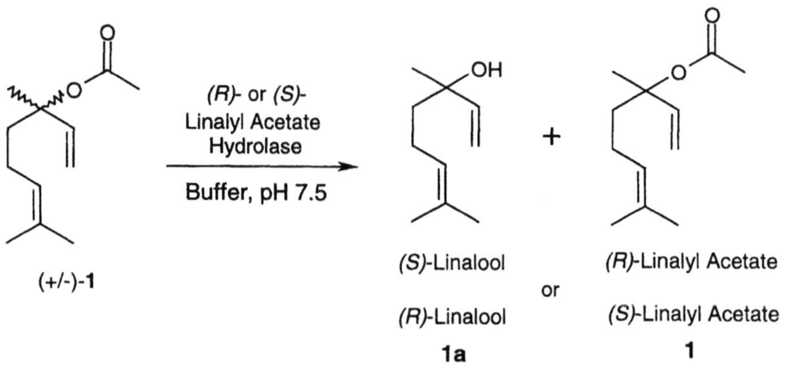

Scheme 1. Biocatalytic resolution of (±)-linalyl acetate

formed from racemic linalyl acetate, its optical purity was exceedingly low (~8%). In addition, significant amounts of α-terpineol and the (non-chiral) terpene alcohols geraniol and nerol were isolated due to undesired allylic rearrangement. On the contrary, various bacterial cells of the genus *Actinomyces* have shown clean hydrolysis without formation of by-products, but the enantioselectivities remained too low to be of practical value ($E<5$) [4]. Most recently, an interesting study has proposed the use of a spacer moiety which is introduced between the sterically offending chiral center and the ester moiety to be hydrolyzed [9]. As a result, the carbonyl moiety of pivaloyloxymethyl- and *n*-butanoyloxymethyl derivatives of linalool became more accessible, making these substrates acceptable for several standard lipases, esterases, and proteases. Although reasonable reaction rates were achieved, the selectivities were not optimal. Best results were obtained with *Candida rugosa* lipase for the corresponding butanoyloxymethyl derivative of linalool ($E=9.7$–9.9).

The reasons for the low enantioselectivities obtained so far may be inherent to the substrate, *i.e.* due to the steric similarity of the methyl and the ethenyl group of linalool, which impedes the chiral recognition process. Alternatively, low selectivities may result from the presence of several competing carboxyl esterases of opposite enantiopreference. This phenomenon is not uncommon in biotransformations; it has been observed during the microbial asymmetric reduction of ketones [10] and β-ketoesters [11], in the enzymatic oxidation of *sec*-alcohols [12], and in *Baeyer-Villiger* oxidations of ketones [13]. This latter option encouraged us to study the enzymes involved in more detail.

Results and Discussion

By extensive screening [4], the best activity on (±)-linalyl acetate was found in cells of *Rhodococcus ruber* DSM 43338; however, the selectivity displayed by whole cells was disappointingly low ($E=4.2$). Two strategies were chosen:

(*i*) Protein purification employing FPLC on hydrophobic interaction material in the first step, following by anion exchange chromatography, revealed that the low selectivities displayed by whole cells are indeed due to the presence of several competing carboxyl esterases. Thus, two active (partially purified) fractions were obtained which showed opposite enantiopreference. However, the low overall yield and the instability of the semipurified enzyme preparations prevented their purification to homogeneity and characterization. However, all of the active enzymes were found to be located in the cytosol, and from SDS-PAGE gels using partially purified preparations it was concluded that the molecular weight was in the range of approximately 30 kDa.

(*ii*) As an alternative approach to protein purification, selective protein extraction experiments based on *PEG*-dextran systems were carried out. This technique allows the extraction of proteins with concomitant (partial) purification in a single operation [14]. Undesired by-products, such as nucleic acids, polysaccharides, and colored dyes (*e.g.* carotenoids) arising from the fermentation are removed in a single step [15]. The system was optimized by variation of three parameters, *i.e.* *pH* (6–8), *PEG*-OMe concentration (5–20% (w/w)), and dextran content (6–20% (w/w)). For the analysis of the data, a response surface central composite design was employed [16]. Based on recommendations given in the literature [17], polyethylene glycol methyl ether (*PEG*-OMe) and dextran from *Leuconostoc mesenteroides* showing molecular weights of 5000 and 260.000, respectively, were chosen.

Aiming at the separation of competing enzymes, the system was first optimized with respect to the phase distribution, *i.e.* enzymes extracted into the (upper) *PEG*-OMe and (lower) dextran phase should display opposite enantiopreference. This approach proved to be successful. Modelling of the response surface for both criteria showed that a relative maximum was found for the following conditions: 11% *PEG*-OMe, 12% dextran, $pH = 8.0$. When protein extracts from both phases were used as catalysts, the desired opposite enantiopreference was observed. The extract from the *PEG*-OMe and the dextran phase gave (*S*)-**1a** ($ee = 20\%$) and (*R*)-**1a** (*ee* 48%) from (±)-**1** at 10% conversion. Next, the system was optimized towards a maximum of activity and a minimum of protein concentration in the *PEG*-OMe-phase. Response surface modelling and calculation of the relative optima for both criteria gave the following data set: using a system consisting of 15% *PEG*-OMe and 6% dextran at *pH* 6.0, 18% of the total protein containing 88% of the activity were located in the *PEG*-OMe-phase. The latter was then subjected to protein purification as described above. The properties of the enzyme preparations thus obtained were very similar to those from the crude cell-free extract by protein chromatography, *i.e.* the selectivity of the enzyme(s) remained intact.

Table 1. Selectivities of linalyl acetate hydrolases from *Rhodococcus ruber* DSM 43338

	(*S*)-enzyme[a] ee_p %[b]		(*R*)-enzyme[a] ee_p %[b]	
	HIC[c]	IEC[c]	HIC[c]	IEC[c]
Crude cell-free extract	<99	n.d.[d]	<99	<99
PEG-OMe/dextran extract	n.d.[d]	n.d.[d]	<99	<99[e]

[a] (*S*)- or (*R*)-linalool were formed from (±)-linalyl acetate, respectively; [b]enantiomeric excess of linalool formed; [c]HIC = hydrophobic interaction chromatography, IEC = ion exchange chromatography; [d]not determined due to loss of activity; [e]the selectivity corresponds to $E = 109$

In order to overcome the limited hydrolytic activity, several attempts towards enzyme induction were carried out. It was found that the presence of competing enzymes possessing different or (almost) no stereoselectivity varied depending on the fermentation conditions. In the absence of any inducer, the cells had hardly any activity at all on (±)-**1**. Thus, *Rhodococcus ruber* DSM 43338 was first grown on a medium composed of glucose and peptone. After reaching the stationary phase, an inducer was added to the culture. Quite surprisingly, the addition of either racemic or (*R*)-(−)-linalyl acetate caused no induction of an enzyme. On the contrary, several carboxyl ester hydrolases capable of accepting linalyl acetate were obtained by using *tert*-butyl acetate as inducer:

(*i*) A highly active enzyme was strongly expressed towards the end of the logarithmic phase of the growth. Unfortunately, it was shown to be non-selective towards (±)-linalyl acetate.
(*ii*) In addition, two enzymes possessing opposite stereopreference were formed when the cells were harvested during the stationary phase, *i.e.* an (*R*)- and an (*S*)-linalyl acetate hydrolase.

Both of these proteins differ significantly in their stability. The (*R*)-selective enzyme proved to be reasonably stable, as only a minor loss of activity was observed upon chromatography, lyophilisation, and freezing at −25°C. Since the addition of *EDTA* had no effect on the activity, the protein was assumed to be devoid of metal ions. On the contrary, (*S*)-linalyl acetate hydrolase was found to be highly sensitive and easily deactivated during purification. Furthermore, a certain degree of deactivation was also observed during protein-selective extraction using the *PEG*-OMe/dextran system.

Experimental

Materials

(±)-Linalyl acetate (Fluka) was purified by silica gel chromatography using petroleum ether (b.p. 60–90°C)/ethyl acetate (20/1) as eluent, followed by *Kugelrohr* distillation (b.p.: 115°C/25 mm). (*R*)-(−)-Linalool (Fluka) was used as received. Polyethylene glycol methyl ether (*MW* = 5000) and dextran from *Leucomostoc mesenteroides* strain no. B-512 (MW 260.000, industrial grade) were obtained from Sigma.

TLC analysis

Samples were analyzed by TLC using Merck silica gel 60 F_{254} and cyclohexane: ethyl acetate = 3:1 as eluent. Compounds were visualized by spraying with vanillin/H_2SO_4 conc. and heat treatment.

GLC analysis

The extent of conversion of the reactions and the enantiomeric excess of linalool was determined *via* GLC analysis using a Varian gas chromatograph 3800 equipped with FID and split/splitless injector using a Chirasil-DEX CB column (25 m × 0.32 mm × 0.25 µm film, Chrompack) and H_2 as carrier gas at 87°C (isothermal). Retention times: (*R*)-linalool, 11.6 min; (*S*)-linalool, 12.0 min; (±)-linalyl acetate, 12.6 min.

Medium and growth conditions

Cells of *Rhodococcus ruber* DSM 43338 were grown in baffled *Erlenmeyer* flasks at 30°C and 130 rpm using the following medium: 10 g/dm^3 glucose, 10 g/dm^3 peptone, 10 g/dm^3 yeast extract, 2 g/dm^3 NaCl, 0.15 g/dm^3 $MgSO_4 \cdot 7H_2O$, 1.3 g/dm^3 NaH_2PO_4 and 4.4 g/dm^3 K_2HPO_4. Enzyme induction was achieved by addition of 4 ml/dm^3 *tert*-butyl acetate or 0.4 ml/dm^3 (*R*)-(−)-linalyl acetate, respectively. The growth of cells was monitored by measurement of the optical density *via* the absorption at 600 nm.

Disruption of cells

To a suspension of *R. ruber* cells (*ca.* 26 g wet weight) in 120 cm^3 of tris-buffer (10 m*M*, *pH* = 7.5), 120 cm^3 of glass beads (diameter 0.35 mm) were added. The cells were disrupted using a Vibrogen cell mill with external cooling (*ca.*+4°C) and 3 shaking cycles of 2 min agitation/5 min cooling each. After removal of the beads by filtration, the crude cell lysate was centrifuged at +4°C and 38000 g.

Protein extraction

To 50 cm^3 of cell-free extract prepared as described above (*pH* adjusted to 6.0), 9.5 g of *PEG*-OMe and 3.8 g dextran were added. The mixture was shaken at 150 rpm at 20°C for 4 h. The phases were then separated by centrifugation (1000 g, 10 min) and directly used for further experiments.

Protein purification

The cell-free extract was loaded directly onto a phenyl sepharose high performance column (d = 16 mm, h = 20 cm, total volume 8 cm^3, Pharmacia). The column was pre-equilibrated with 10 m*M* tris containing 1 *M* of $(NH_4)_2SO_4$ (*pH* = 7.5) at a flow rate of 3 cm^3/min. Elution was effected using a stepwise gradient with 10 m*M* tris (*pH* = 7.5) at intervals of 100/75/50/25/0%. The (*S*)- and the non-selective enzyme were desorbed with 100% 10 m*M* tris (*pH* = 7.5), whereas the (*R*)-enzyme proved to be more lipophilic as it eluted with 100% destilled H_2O. Active fractions were subjected to further purification by ion exchange chromatography using a Q6 column (total volume 6 cm^3, BioRad), pre-equilibrated with 10 m*M* tris-buffer (*pH* = 7.5). Elution was performed with 10 m*M* tris-buffer containing 1 *M* NaCl (*pH* = 7.5, buffer B) with a linear gradient from 0–1 *M* NaCl at a flow rate of 6 cm^3/min. Desorption of (*R*)- and (*S*)-enzyme took place in the range of 13.4–19.4% and 25.3–27.2% of buffer B, respectively. For the SDS-PAGE experiments the Pharmacia Phastsystem equipment was used.

Enzyme assay

Samples of the eluent (6 mm^3 total volume) were mixed with 10 mm^3 (±)-linalyl acetate on a vortex mixer and incubated for 16 h at room temperature under gentle rolling. The reaction was quenched by addition of 1 cm^3 of ethyl acetate. The samples were extracted four times with 4 cm^3 ethyl acetate each. The organic phase was dried over Na_2SO_2 and concentrated by evaporation. The samples were analyzed by GLC as described above. The absolute configurations of **1** and **1a** was determined by chiral GLC *via* co-injection with commerically available samples.

Acknowledgements

This work was performed within the SFB *Biokatalyse*; financial support by the Austrian Science Foundation (project no. F115) is gratefully acknowledged.

References

[1] Ohloff G, Klein E (1962) Tetrahedron **18**: 37
[2] Okumura S, Iwai M, Tsujisaka Y (1979) Biochim Biophys Acta **575**: 156
[3] Iwai M, Okumura S, Tsujisaka Y (1980) Agric Biol Chem **44**: 2731
[4] Osprian I, Steinreiber A, Mischitz M, Faber K (1996) Biotechnol Lett **18**: 1331a
[5] For a rare exception see: Schultz M, Hermann P, Kunz H (1992) Synlett 37
[6] For a review on the biocatalytic resolution of carboxylic esters containing fully substituted chiral carbon atoms see: Pogorevc M, Faber K (1999) J Mol Catal B Enzymatic (in press)
[7] Madyastha KM, Krishna Murthy NSR (1988) Appl Microbiol Biotechnol **28**: 324
[8] Oritani T, Yamashita K (1973) Agric Biol Chem **37**: 1923
[9] Franssen MCR, Goetheer ELV, Jongejan H, de Groot A (1998) Tetrahedron Lett **39**: 8345
[10] Sih CJ, Shieh W-R, Chien C-S, Wu S-H, Girdaukas G (1986) Ann NY Acad Sci **471**: 239
[11] Nakamura K, Matsuda T, Itoh T, Ohno A (1996) Tetrahedron Lett **37**: 5727
[12] Barrett CH, Dodgson KS, White GF, Payne WJ (1980) Biochem J **187**: 703
[13] Grogan G, Roberts S, Wan P, Willetts A (1993) Biotechnol Lett **15**: 913
[14] Verman J, Kula MR (1990) Biotechnol Appl Biochem **12**: 397
[15] Kroner KH, Hustedt H, Kula M-R (1982) Biotechnol Bioeng **24**: 1015
[16] Box GEP, Hunger WG, Hunter JS (1978) Statistics for Experiments. An Introduction to Design, Data Analysis and Model Building. Wiley, New York
[17] Walter H, Johansson G (1986) Anal Biochem **155**: 215

Received November 18, 1999. Accepted January 21, 2000

Microbial and Enzymatic Synthesis of Optically Pure *D*- and *L*-3-Trimethylsilyl-alanine by Deracemization of *D,L*-5-Trimethylsilyl-methyl-hydantoin

Markus Pietzsch[1,*], **Thomas Waniek**[1], **Richard J. Smith**[2], **Svetoslav Bratovanov**[2], **Stefan Bienz**[2], and **Christoph Syldatk**[1]

[1] Institute of Biochemical Engineering, University of Stuttgart, D-70569 Stuttgart, Germany
[2] Institute of Organic Chemistry, University of Zurich, CH-8057 Zurich, Switzerland

Summary. The sterospecificities of hydantoinases and N-carbamoyl amino acid amidohydrolases (N-carbamoylases) from different microbial sources were investigated for the stereoselective syntheses of the unnatural silicon-containing amino acids *D*- and *L*-3-trimethylsilyl-alanine (**3**) from the respective racemic hydantoin *D,L*-**1**. In a preparative biotransformation, whole resting cells of *Agrobacterium sp.* IP I 671, immobilized in a Ca-alginate matrix, were used for the synthesis of amino acid *D*-**3** in 88% yield and 95% enantiomeric excess. Since the purified *D*-N-carbamoylase from *Agrobacterium sp.* IP I 671 was shown to be 100% *D*-selective, the enantiomeric purity of 95% of *D*-**3** arising from the transformation with the immobilized cells must be explained by the participation of a further, *L*-selective N-carbamoylase or, which is more likely, by racemization of the final hydrolysis product by the action of an amino acid racemase. Isolated hydantoinases from *Bacillus thermoglucosidasius*, *Thermus sp.*, *Arthrobacter aurescens* DSM 3745, and *Arthrobacter crystallopoietes* DSM 20117 turned out to be stereospecific for the conversion of the *D*-form of hydantoin *D, L*-**1**. The enantiomerically pure *L*-form of **3** was also prepared. It was synthesized from racemic N-carbamoyl amino acid *D,L*-**2** by enantiomer-specific hydrolysis of the *L*-form in presence of *L*-N-carbamoylase from *Arthrobacter aurescens* DSM 3747.

Keywords. *D*-3-Trimethylsilyl-alanine; Hydantoinase; N-Carbamoyl amino acid amidohydrolase; N-Carbamoylase; Biotransformation.

Introduction

Optically active silicon-containing amino acids represent unnatural compounds with interesting pharmaceutical behavior. When, e.g., in a biological active renin inhibitor *L*-phenylalanine was replaced by *L*-3-trimethylsilyl-alanine (*L*-**3**), in contrast to the parent compound the silicon-containing analog was largely resistant towards proteolytic digestion, whereas the inhibition constant was affected only marginally [1].

* Corresponding author

Several approaches towards optically active silicon-containing amino acids have been published. The *Schöllkopf*, *Seebach*, or *Myers* methods of asymmetric α-alkylation of glycine enolate equivalents [1–3] or the *Evans* procedure of asymmetric α-bromination of carboxylic acid derivatives followed by substitution of the bromine atom [4] gave access – at least principally – to either antipode of compound **3** in high enantiomeric excesses. The overall yields attained with these syntheses, however, were rather low, and chiral auxiliaries had to be used in stoichiometric amounts.

In 1996, *Yamananka et al.* reported on the first biotransformation for the preparation of a silicon-containing amino acid. With *L*-aminoacylase from porcine liver as the biocatalyst, *L*-**3** was prepared by enantiomer-specific hydrolysis of *L*-N-acetyl-3-trimethylsilyl-alanine from the racemate [5]. Since this reaction represents a kinetic resolution, a yield of 50% can be obtained at its best, and the transformation is thus not efficient. The same drawback suffers the kinetic resolution of a silicon-containing N-carbamoyl amino acid (synthesis of *D*-*p*-trimethylsilylphenyl-alanine) by the action of a *D*-N-carbamoylase from *Blastobacter sp.* A17 p-4 [6].

If applicable to silicon-containing substrates, no limitation with respect to yields would be present in the hydantoinase process that was studied and developped in our group [7]. The hydantoinase process consists of a casade of two hydrolytic enzymes *i.e.* a hydantoinase and a N-carbamoylase (see Scheme 1), in combination with chemical or enzymatic racemization of the starting hydantoin. The hydantoinase is responsible for the partial hydrolysis of a hydantoin to the respective N-carbamoyl amino acid and the N-carbamoylase for the subsequent deaminocarbonylation to the corresponding amino acid. Because 5-monosubstituted hydantoins racemize spontaneously already under slightly alkaline conditions, enzymatically accepted enantiomeric form of the hydantoin is constantly resupplied from the racemate. A theoretical yield of 100% is thus possible for the formation of enantiomerically pure amino acids departing from racemic starting materials.

We describe in the following that silicon-containing amino acids are in fact accessible by the hydantoinase process. Several biocatalysts of different microorganisms have been studied and were found to hydrolyze enantio-selectively one component of *D,L*-5-trimethylsilylmethyl-hydantoin (*D,L*-**1**) or *D,L*-N-carbamoyl-3-trimethylsilyl-alanine (*D,L*-**2**), respectively. The biocatalysts investigated are whole resting cells of *Agrobacterium sp.* IP I 671 and its enriched N-carbamoylase [8], commercially available hydantoinase from *Bacillus thermoglucosidasius* and

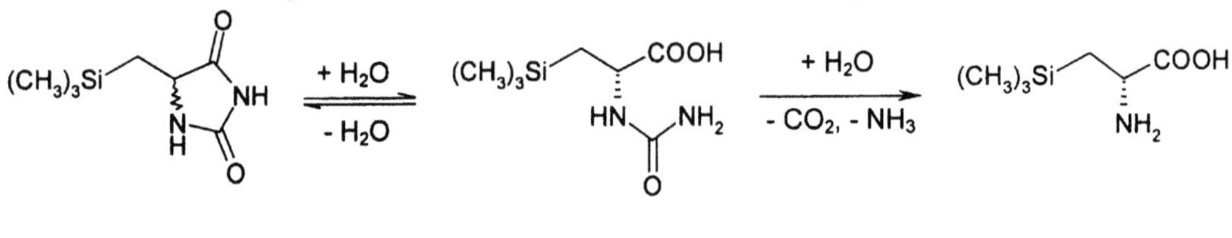

1 **2** **3**

Scheme 1

Thermus sp., purified hydantoinase from *Arthrobacter aurescens* DSM 3745 [9], and recombinant *L*-N-carbamoylase from *Arthrobacter aurescens* DSM 3747 [10].

Results and Discussion

Biotransformations with whole cells of Agrobacterium sp. IP I 671

It has been shown by *Runser et al.* in 1990 that whole cells of *Agrobacterium sp.* IP I 671 catalyze the hydrolysis of *D,L*-5-(*p*-hydroxyphenyl)-hydantoin to form enantiomerically pure *D*-(*p*-hydroxyphenyl)-glycine in excellent yield [11]. We have tested the same system for the biotransformation of our silicon-containing test substrate, hydantoin *D,L*-**1**. Hence, racemic hydantoin *D,L*-**1** was incubated with immobilized cells of *Agrobacterium sp.* IP I 671 (25% cell wet mass (CWM) in Ca-alginate) under anaerobic conditions and delivered the *D*-form of the desired amino acid in 88% yield and 95% *ee*. The kinetics of the transformation, together with the mass balance, are graphically shown in Fig. 1a.

It is imperative for the transformation that it is carried out under anaerobic conditions. In presence of oxygen, partial decomposition of the desired amino acid is observed. Figure 1b shows that the curve for the mass balance (of substrate, intermediate, and product) declines rapidly up to a reaction time of approximately 40 h. After this time, the concentration of the amino acid as well as the mass balance remain almost constant.

The degradation of the amino acid might be explained by the action of an amino acid dehydrogenase. In *Agrobacterium sp.* IP I 671, the gene coding for an enzyme with close homologies to such amino acid dehydrogenases was found to be located on the same operon as the hydantoinase and the N-carbamoyl amino acid amidohydrolase [12]. Evidently, the enzyme responsible for the degradation reaction is deactivated after approximately 40 h, not affecting the transformation any longer.

The silicon-containing amino acid *D*-**3**, formed in a preparative biotransformation of 500 mg of hydantoin *D,L*-**1**, was isolated and purified by means of ion exchange chromatography. The amino acid had to be separated from a number of other compounds that were liberated from the whole cell biocatalyst. As is recognized from the reversed phase HPLC chromatograms shown in Fig. 2, several compounds that are not related to the silicon-containing substrate were found in the broth of both the biotransformation and a substrate-free incubation of the cells in the buffer.

The enantiomeric excess of the *D*-3-trimethylsily-alanine (*D*-**3**) obtained by the hydantoinase process with whole cells of *Agrobacterium sp.* IP I 671 was determined to be 95%. Since isolated *D*-N-carbamoylase from the same microorganism turned out to be 100% *D*-selective (see below), not delivering traces of the *L*-antipode of **3**, the lower enantiomeric purity obtained by the biotransformation in presence of the whole cells was rather surprising. It can be explained, however, with the possible participation of a further, *L*-selective N-carbamoylase – as shown to be present, *e.g.*, in a *Blastobacter* strain [6] – or with the more likely possibility of racemization of *D*-3-trimethylsilyl-alanine (*D*-**3**) by the action of an amino acid racemase [13]. Altogether, several of the drawbacks commonly associated with the use of whole cell

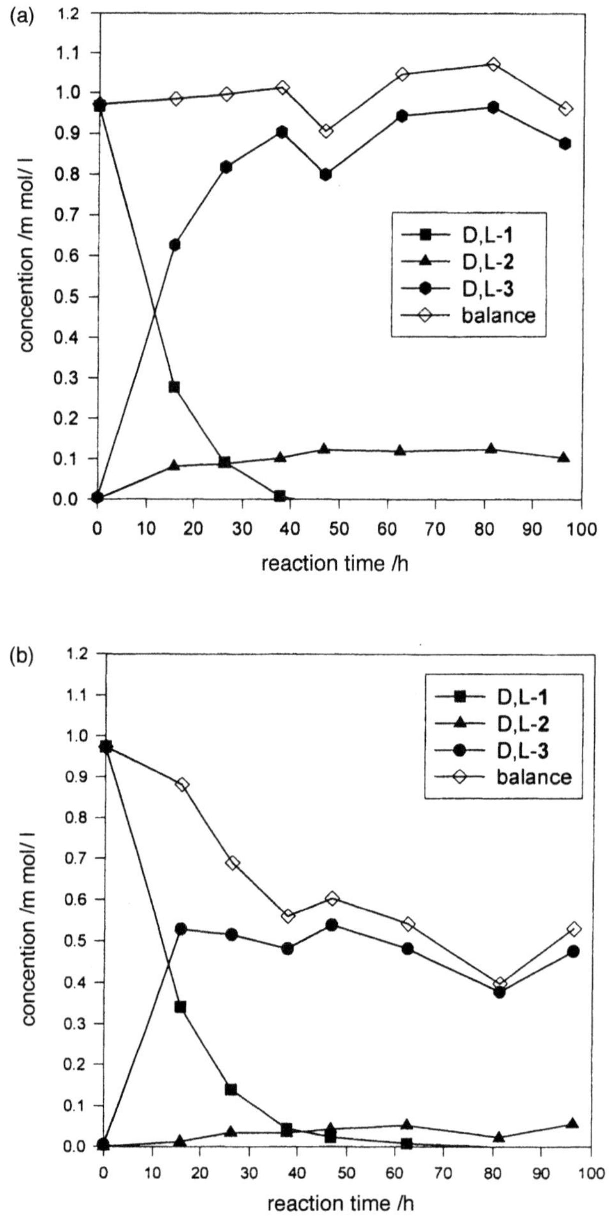

Fig. 1. Hydrolysis of *D,L*-**1** with immobilized cells of *Agrobacterium sp.* IP I 671 as the biocatalyst under (a) anaerobic and (b) aerobic conditions

biocatalysts [14] were experienced with the whole cell biotransformation of *D,L*-**1**, too. Therefore, several isolated enzymes, either in free or in immobilized form, were investigated for their ability to enantio-selectively hydrolyze racemic hydantoin *D,L*-**1** or racemic N-carbamoyl amino acid *D,L*-**2**.

*Hydrolysis of racemic hydantoin D,L-***1** *with hydantoinases*

Racemic hydantoin *D,L*-**1** was treated with H_2O in the presence of immobilized hydantoinase from *Arthrobacter crystallopoietes* DSM 20117 (on Eupergit C,

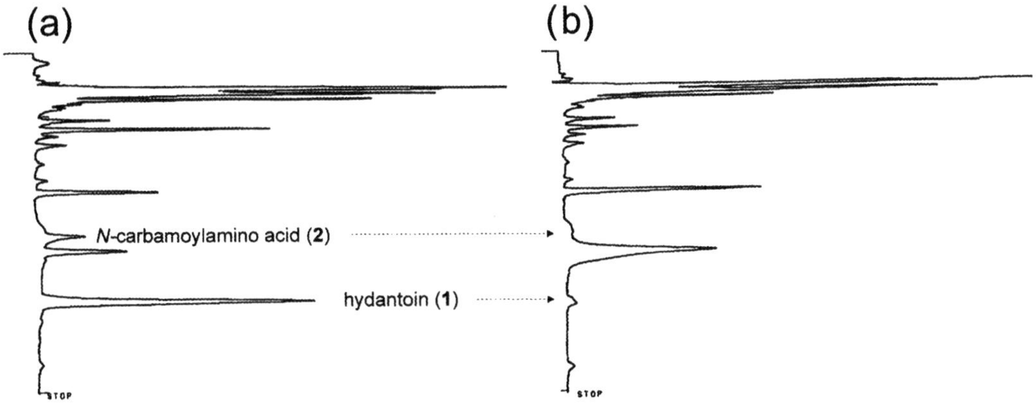

Fig. 2. (a) HPLC chromatogram of a sample taken during the biotransformation of *D,L*-**1** in presence of immobilized cells of *Agrobacterium sp.* IP I 671 and (b) HPLC chromatogram of a reference sample of the broth of the same cells without substrate

[15]), free hydantoinase from *Arthrobacter aurescens* DSM 3745 [9], and the commercially available hydantoinases from *Bacillus thermoglucosidasius* and *Thermus sp.* All four enzymes were found to catalyze the partial hydrolysis of *D,L*-**1** to the N-carbamoyl amino acid **2**. As disclosed by HPLC on a chiral medium, all enzymes were stereospecific for the *D*-enantiomer, delivering exclusively the *D*-form of compound **2**. The stereochemical results of the reactions, except for the transformation in presence of the hydantoinase from *Arthrobacter aurescens* DSM 3745, correspond to our expectations, which are based on the enzyme-catalyzed transformations of other hydantoin derivatives. The hydantoinase from *Arthrobacter aurescens* DSM 3745 represents a special case. It is known to change its enantiomeric preference with the substrate, converting, *e.g.*, the hydantoins of tryptophan [9] and phenylalanine [15] with *L*-selectivity and the hydantoin of methionine with *D*-selectivity [9]. The silicon-containing hydantoin **1** is the first substrate which is converted by *Arthrobacter aurescens* DSM 3745 with complete *D*-stereospecificity.

*Hydrolysis of racemic N-carbamoyl amino acid D,L-**2** with free and immobilized D- and L-N-carbamoylases*

Since hydantoinases catalyze the reversible hydrolysis of hydantoins to N-carbamoyl amino acids, most related biotransformations deliver mixtures of substrates and products. In the case of silicon-containing hydantoin *D,L*-**1**, an equilibrium mixture of *D,L*-**1** and *D*-**2** in a ratio of approximately 20:80 was formed (corresponds to a conversion of about 80%). To shift the equilibrium towards the side of the product, an N-carbamoylase with matching stereospecifity should be added to the system. Thus, two N-carbamoylases, catalyzing the irreversible deaminocarbonylation of N-carbamoyl amino acids to the free amino acids, were investigated for their substrate- and stereoselectivity. Enriched *D*-N-carbamoylase from *Agrobacterium sp.* IP I 671, used in free form, catalyzed the hydrolysis of the *D*-enantiomer of racemic N-carbamoyl-3-trimethylsilyl-alanine *D,L*-**2**. As can be seen from Fig. 3, the enzyme-

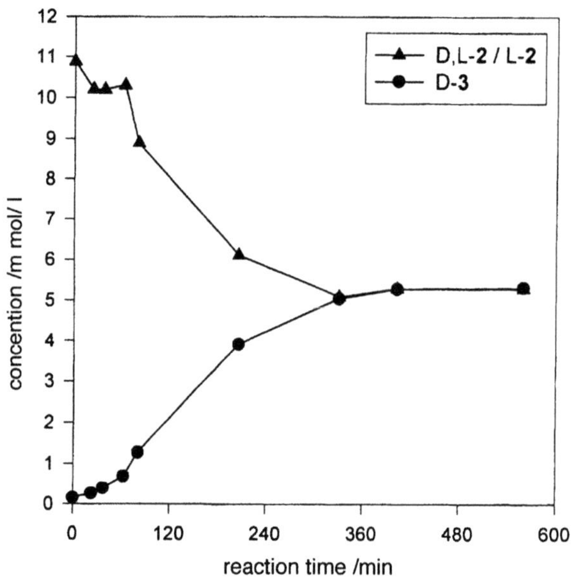

Fig. 3. Hydrolysis of *D,L*-N-carbamoyl-3-trimethylsilyl-alanine (*D,L*-**2**) with enriched *D*-N-carbamoylase from *Agrobacterium sp.* IP I 671 as the biocatalyst

catalyzed hydrolysis leads to the expected 50% conversion only, and it was shown by HPLC that exclusively *D*-3-trimethylsilyl-alanine (*D*-**3**) was formed in the process. On the other hand, the hydrolysis of *D,L*-**2** by the action of the N-carbamoylase from *Arthrobacter aurescens* DSM 3747 (immobilized enzyme), which is known for its *L*-stereospecifity, delivered exclusively the *L*-enantiomer of **3** after 50% conversion.

In summary, the hydantoinase process, an elegant method for the synthesis of enantiomerically pure unnatural amino acids, proved applicable for the production of enantiomerically pure *D*-3-trimethylsilyl-alanine (*D*-**3**), too. Four isolated hydantoinases, two N-carbamoylases, and a whole cell biocatalyst were tested as biocatalysts. They all proved to accept the silicon-containing substrates, and the biotransformations led with high stereoselectivities to the respective hydrolysis products. It is currently under investigation, whether other 5-monosubstituted silicon-containing hydantoins and N-carbamoyl amino acids are stereospecifically converted by these enzymes as well.

Experimental

General

Unless otherwise stated, all chemicals were of reagent grade and purchased from Fluka Chemie AG. Aqueous solutions were prepared with ultrapure H_2O (NANOpure II, Fa. Barnstedt, Newton, MA, USA). *D,L*-5-Trimethylsilylmethyl-imidazolidin-2,4-dione (*D,L*-5-trimethylsilymethyl-hydantoin, *D,L*-**1**), *D,L*-N-carbamoyl-3-trimethylsilyl-alanine (*D,L*-**2**), and *D,L*-**3**-trimethylsilyl-alanine (*D,L*-**3**) were synthesized as described previously [17, 18]. Since the non SI-unit U (μmol $\cdot$ min^{-1}) is commonly used in the literature, this unit was also used in the present contribution. 1 U corresponds to 16.67 nKat (10^{-9}mol $\cdot$ s^{-1}).

Microorganisms

Agrobacterium sp. IP I 671 was obtained from the Institute Pasteur (Paris, France) and cultivated according to *Meyer* and *Runser* [19]. After centrifugation, the whole cells were immobilized by entrapment in Ca-alginate as described previously for cells of *Trigonopsis variabilis* [20] with the following modifications: the cell wet mass (50 g) was suspended in 25 cm^3 of a 0.9% NaCl solution and mixed with 204 g of a solution of 4 g Na-alginate in 200 cm^3 of H_2O.

Enzymes

Enriched *D*-N-carbamoylase from *Agrobacterium sp.* IP I 671 and hydantoinases from *Bacillus thermoglucosidasius* (Hyd 1, carrier-fixed) and *Thermus sp.* (Hyd 2, suspension) were gifts from Roche Diagnostics (Penzberg, Germany). Hydantoinase from *Arthrobacter aurescens* DSM 3745 was purified as described previously [9]. Hydantoinase from *Arthrobacter crystallopoietes* DSM 20117, purified according to the literature [21], was immobilized on Eupergit C (Röhm Pharma, Germany) [15]. *L*-N-Carbamoylase from *Arthrobacter aurescens* DSM 3747 was purified from recombinant *Escherichia coli* [22] and immobilized on EAH-Sepharose [23].

Analytical methods

RP-HPLC: The concentrations of **1** and **2** were determined with reversed phase HPLC on a Gromsil ODS 1 PE column (5 μm, 250×4,6 mm, Grom, Herrenberg, Germany; eluent: 7 cm^3 of aq. H_3PO_4 (85%) in 2000 cm^3 of H_2O and 500 cm^3 of MeOH; flow rate: 1.0 $cm^3\ min^{-1}$) according to Ref. [9] with UV-detection at 210 nm. Retention times: **1**: 35.2 min; **2**: 28.0 min.

Quantification of **3**: Concentrations of **3** were measured photometrically after reaction of the amino acid with ninhydrin [24]. A 100 mm^3 sample (concentration of amino acid between 0.1 and 1 m*M*) was mixed with 100 mm^3 of acetate buffer (21.33 g of AcONa and 6.9 cm^3 of AcOH in 100 cm^3 of H_2O) and kept at 60°C for 5 min. To this mixture, 100 mm^3 of a ninhydrin solution (174 mg of ninhydrin and 174 mg of hydrindantin×2 H_2O in 15 cm^3 of 2-methoxyethanol) were added, and the sample was kept at 60° for 20 min. Afer cooling to 23°C it was diluted with *i*-PrOH/H_2O (1:1, 1000 mm^3), and the extinction was measured at a $\lambda = 570$ nm with a spectrophotometer (Ultrospec III, Pharmacia, Freiburg, Germany). Calibration was done with *D,L*-**3** in concentrations between 0.1 and 1 m*M*.

HPLC on a chiral phase: The separation of the enantiomers of *D,L*-**1** and *D,L*-**2** was accomplished by HPLC on an ET 200/4 Nucleodex β-PM column (Macherey-Nagel, Düren, Germany; eluent: 7 cm^3 H_3PO_4 (85%) in 2000 cm^3 H_2O and 500 cm^3 MeOH, *pH* adjusted to 3.7 with conc. aq. NaOH solution; flow rate: 0.2 $cm^3 \cdot min^{-1}$) coupled to a Spectra System (Thermo Separation Products, Egelsbach, Germany) with UV detection at 210 nm. Retention times: (*L*)-**1**: 77.0 min, (*D*)-**1**: 92.8 min, $\alpha = 1.21$, (*L*)-**2**: 28.2 min, (*D*)-**2**: 25.6 min, $\alpha = 1.10$.

*Bioconversion of D,L-**1** with immobilized whole cells of Agrobacterium sp. IP I 671*

Analytical scale: Under a blanket of N_2, racemic hydantoin *D,L*-**1** (9.3 mg, 50 μmol) was suspended in 50 cm^3 of a 0.1 *M* Tris/HCl-buffer containing 1% of $CaCl_2$ (*pH* 8.5), and immobilized cells of *Agrobacterium sp.* IP I 671 (5 g) were added at 37°C (a reference reaction was carried out without the N_2 atmosphere). The mixture was kept at 37°C, and samples were taken at intervals of 10–20 h and analyzed to determine the concentrations of **1**, **2**, and **3** as described above.

Preparative scale: Since large amounts of Tris/HCl buffer and $CaCl_2$ are difficult to remove from the reaction products the preparative scale bioconversion was carried out in pure H_2O. Under a blanket of N_2, racemic hydantoin *D,L*-**1** (460 mg, 2.5 mmol) was suspended in H_2O (200 cm^3). Immobilized cells (30 g) were added, the mixture was kept at 37°C, and samples were drawn and analyzed in regular intervals to follow the reaction. After complete consumption of *D,L*-**1** (26 h), the immobilizate was filtered off, the filtrate passed through a membrane filter (0.45 μm)., and concentrated to 35 cm^3 by rotary evaporation. A portion of the concentrate (25 cm^3) was applied with a flow rate of 0.25 $cm^3 \cdot min^{-1}$ to a chromatography column (XK16/20, packed with 20 g of Dowex 50 WX8, 400 mesh, H^+) connected to an FPLC-system equipped with a UV spectrometer and a conductivity cell (Pharmacia, Freiburg, Germany). The column was washed with H_2O until absorption and conductivity remained constant before the amino acid was eluted by washing with 0.5 *M* NH_4OH at a flow rate of 10 $cm^3 \cdot min^{-1}$. Samples containing amino acid were detected by reaction with ninhydrin on a TLC plate, collected, concentrated at a temperature of < 50°C, and dried at 10^{-3} torr to yield 242 mg (1.5 mmol, 88%) of *D*-**3** (95% *ee*, see below) which was virtually pure (^{1}H NMR).

To obtain a sample for analysis, 28 mg (0.17 mmol) of *D*-**3** were dissolved in conc. aq. HCl, and the solvent was removed by evaporation. A solution of 0.16 cm^3 (0.51 mmol) of trimethylsilyl-diethylamine in 3 cm^3 of *THF* was added, and the mixture was stirred for 1 h. The precipitate was filtered off and washed with Et_2O (8 cm^3), the combined organic phases were concentrated, and the residue was dried at 10^{-3} torr. Hydrolysis was effected by the addition of acetone (1 cm^3) and H_2O (4 cm^3) and stirring for 30 min. The solvent was evaporated, the solid residue washed with acetone (2 cm^3), MeOH (1 cm^3), and Et_2O (4 cm^3), and dried at 10^{-3} torr to give 12.3 mg (45%) of *D*-**3** (95% *ee*).

$[\alpha]_D^{25} = -26.4$ ($c = 0.22$, 2 *M* aq. HCl) literature values for *L*-**3**: +31.0 ($c = 0.51$, 4 M aq. HCl) [2], +33.3 ($c = 0.526$, 4 *M* aq. HCl) [3], +21.4 ($c = 3.2$, aq. HCl) [4].

To determine the enantiomeric purity of the sample, *D*-**3** was transformed into *D*-**1** by reaction with KOCN and HCl [25] and analyzed by HPLC on a chiral phase. The analytical data of *D*-**3** (^{1}H and ^{13}C NMR spectroscopy elementary analysis, MS) were identical with those obtained for the racemic amino acid from which *D,L*-**1** was originally synthesized.

*Biotransformations of D,L-**1** with free and immobilized hydantoinases*

Hydantoinase from *Arthrobacter aurescens* DSM 3745 (31 mU; activity for the conversion of the standard *D,L*-5-benzylhydantoin) was added to 1 cm^3 of a 1 m*M* solution of *D,L*-**1** in 0.1 *M* Tris/HCl-buffer (*pH* = 8.5) at 37°C, and the mixture was kept at this temperature for the transformation. At a conversion of 8%, attained after several hours, 100 mm^3 of the mixture were diluted with 900 mm^3 of HPLC eluent, and, after centrifugation, the concentrations of substrate and product enantiomers were analyzed by chiral HPLC. Analogously, 0.5 cm^3 of a 1 m*M* soln. of *D,L*-**1** in 0.1 *M* Tris/HCl-buffer (*pH* 8.5) were treated with the hydantoinase from *Thermus sp.* (10 mm^3, 120 U; activity for the conversion of *D,L*-5-phenylhydantoin), the hydantoinase from *Bacillus thermoglucosidasius* (30 mg, 6.8 U; activity for the conversion of *D,L*-5-(*p*-hydroxyphenyl-hydantion), and the hydantoinase from *Arthrobacter crystallopoietes* DSM 20117 (70 mg, 0.1 U; activity for the conversion of *D,L*-5-(*p*-hydroxyphenyl)-hydantoin), respectively. At 3%, 16%, and 6% conversion, attained after several hours of reaction at 37°C, 100 mm^3 of the mixtures were diluted with 900 mm^3 of HPLC eluent and analyzed as described above by chiral HPLC.

*Biotransformations of D,L-**2** with free N-carbamoylases*

Racemic N-carbamoyl amino acid *D,L*-**2** (20 mg, 0.1 mmol) was dissolved in 10 cm^3 of 0.1 *M* Tris/HCl-buffer (*pH* 8.5), lyophilized *D*-N-carbamoylase from *Agrobacterium sp.* IP I 671 (4 mg, 0.8 U; activity for the conversion of *D,L*-N-carbamoylphenylalanine) was added, and the mixture was kept at 37°C. At different time intervals (see Fig. 3) samples were taken and analyzed by RP-HPLC and photometrically after reaction with ninhydrin. Chiral HPLC of the samples revealed that only *D*-**2** was hydrolyzed. Analogously, *D,L*-**2** (1000 mm^3 of a 3 mM soln. in 0.1 *M* Tris/HCl-buffer (*pH* 8.5)) was treated with immobilized *L*-N-carbamoylase from *Arthrobacter aurescens* DSM 3747 (59.3 mg) at 37°C. After 2 h, a sample was taken and centrifuged, and the supernatant was analyzed by chiral HPLC to reveal that solely *L*-2 was hydrolyzed.

Acknowledgements

We thank Dr. *W. Tischer* and Dr. *W. Bacher* (Roche Diagnostics, Penzberg, Germany) for the provision of the immobilized hydantoinases from *Bacillus thermoglucosidasius* and *Thermus sp.* and of the enriched *D*-N-carbamoylase from *Agrobacterium sp.* IP I 671.

References

[1] Weidmann B (1992) Chimia **46**: 312
[2] Fitzi R, Seebach D (1988) Tetrahedron **44**: 5277
[3] Myers Ag, Gleason JL, Yoon T, Kung DW (1997) J Am Chem Soc **119**: 656
[4] Walkup RD, Cole DC, Whittlesey BR (1995) J Org Chem **60**: 2630
[5] Yamanaka H, Fukui T, Kawamoto T, Tanaka A (1996) Appl Microbiol Biotechnol **45**: 51
[6] Tsuji Y, Yamanaka H, Fukui T, Kawamoto T, Tanaka A (1997) Appl Microbiol Biotechnol **47**: 114
[7] Pietzsch M, Syldatk C (1995) Hydrolysis and formation of hydantoins. In: Drauz K, Waldmann H (eds) Enzyme Catalysis in Organic Synthesis. VCH-Verlag, Weinheim, p 409
[8] Runser SM, Meyer PC (1993) Eur J Biochem **213**: 1315
[9] May O, Siemann M, Pietzsch M, Kiess M, Mattes R, Syldatk C (1988) J Biotechnol **61**: 1
[10] Wilms B, Wiese A, Syldatk C, Mattes R, Altenbuchner J, Pietzsch M (1999) J Biotechnol **68**: 101
[11] Runser S, Chinski N, Ohleyer E (1990) Appl Microbiol Biotechnol **33**: 382
[12] Hils M (1998) Mutanten der D-Carbamoylase zur Bildung aktiven Enzyms bei Expression des Gens in *Escherichia coli* und Analyse eines Genclusters für die Enzyme des Hydantoin-Abbaus aus *Agrobacterium sp.* IP I-671. PhD Thesis, Institute of Industrial Genetics, Stuttgart, Germany
[13] Möller A, Syldatk C, Schulze M, Wagner F (1988) Enzyme Microb Technol **10**: 618
[14] Faber K (1997), Biotransformations in Organic Chemistry – A Textbook. Springer, Berlin
[15] Pietzsch M, Schwämmle A, Syldatk C (in prep.)
[16] Waniek T (1999) Untersuchungen zur Substratspezifität und Enantioselektivität mikrobieller Hydantoinasen. PhD Thesis, Institute of Biochemical Engineering, Stuttgart, Germany
[17] Porter TH, Shive W (1968) J Med Chem **11**: 402
[18] Smith RJ, Bratovanov S, Bienz S (1997) Tetrahedron **53**: 13695
[19] Meyer P, Runser S (1993) FEMS Microbiol Lett **109**: 67
[20] Huber P, Bratovanov S, Bienz S, Syldatk C, Pietzsch M (1996) Tetrahedron Asymm **7**: 69
[21] Siemann M, Alvarado-Marín Á, Pietzsch M, Syldatk C (1998) J Molec Catal B, Enzymatic **6**: 387
[22] Pietzsch M, Wiese A, Ragnitz K, Wilms B, Altenbuchner J, Mattes R, Syldatk C (1999) J Chromatogr B **737**: 179
[23] Ragnitz K, Syldatk C, Pietzsch M (in prep.)
[24] Fahmy RA, Niederwieser A, Pataki G, Brenner M (1961) Helv Chim Acta **44**: 2022
[25] Dakin HD (1919) Biochem J **13**: 398

Received December 20, 1999. Accepted January 7, 2000

Immobilization of *Rhodococcus* AJ270 and Use of Entrapped Biocatalyst for the Production of Acrylic Acid

John Colby*, **David Snell**[a], and **Gary W. Black**

School of Sciences, University of Sunderland, Sunderland SR1 3SD, UK

Summary. *Rhodococcus* AJ270 is adsorbed by Dowex 1 at 15.4 mg dry weight per g resin with maximum amidase specific activity observed at lower loadings. Bacteria form a monolayer on the resin surface, and adsorption is complete within 2 min. AJ270 can be entrapped in agar and agarose gels (optimum loading: 20 mg dry weight bacteria per cm^3 gel). Adsorption and entrapment improve amidase thermal stability 3–4 fold, and entrapment shifts the *pH* optimum from 8 to 7. Adsorbed and free bacteria show similar values for K_m and V_{max}, but entrapped bacteria have higher K_m values. Compared with bacteria adsorbed to Dowex, the activity per cm^3 of matrix of agar-entrapped AJ270 is eight-fold higher. In stirred-tank reactors, exposure to acrylic acid reduces the amidase activity of the biocatalyst in the hydrolysis of acrylamide. In column reactors, entrapped AJ270 suffers little reduction in amidase activity against 0.25 *M* acrylamide over 22 h continuous operation.

Keywords. Acrylamide; Amidase; Acrylate; Biocatalysis; *Rhodococcus*.

Introduction

When grown on acetamide, *Rhodococcus* AJ270 expresses high specific activities of nitrile hydratase and amidase [1], enzymes with broad specificity against a wide range of nitriles and amides [2,3]. Washed suspensions of AJ270 have been used for the enantioselective hydrolysis of (*R,S*)-2-phenylbutyronitrile [2] and of racemic ibuprofen amide, in the latter case yielding the pharmacologically active non-steroidal anti-inflammatory agent *S*-(+)-ibuprofen in 94% *e.e.* [1]. This versatile biocatalyst also catalyses the regioselective hydrolysis of certain dinitriles [2,4]. These applications used the rhodococcal biocatalyst in a 'one-shot' batch mode, whereas immobilization of the bacteria would allow multiple re-use of the catalyst, possibly in continuous reactors, and would certainly facilitate product recovery. This paper describes methods for the adsorption and entrapment of AJ270, reports changes in the properties of the amidase activity resulting from immobilization, and investigates the use of entrapped bacteria for the hydrolysis of acrylamide. This latter biotransformation is of commercial significance for the

* Corresponding author

[a] Present address: Helena Biosciences, Sunderland Business Park, Sunderland SR5 3XB, UK

Table 1. Adsorption of AJ270 to anionic exchange resins. Bacteria (4.2 mg dry weight) were immobilized to the anionic exchangers (3 g wet weight), and the % of bacteria adsorbed determined as described in the Experimental section. For the data in the second column, the bacteria and resins were washed with buffer containing 0.01 *M EDTA* prior to immobilization

Resin	% of bacteria adsorbed	
	−EDTA	*+EDTA*
Amberlite IRA 904	6	22
Amberlite IRA 400	21	24
Amberlite IRA 402	2	7
Amberlite IRA 440c	6	7
Amberlite IRA 900	4	14
Dowex 1X8-400	96	94
Dowex 2X8-400	60	50

production of acrylates, commodity chemicals produced globally on a large scale for the manufacture of a wide variety of polymers [5,6].

Results and Discussion

Adsorption of AJ270 to anionic exchange resins

Since bacteria generally carry an overall negative charge at physiological *pH*, a variety of anionic exchange resins were tested as adsorption matrices. Most of them except Dowex 1 and 2 showed poor adsorption of AJ270, with Dowex 1 having the highest capacity (Table 1). Treatment of the bacteria with *EDTA* to remove cations from the bacterial surface improved attachment to some of the Amberlite resins significantly, but Dowex 1 remained the best adsorption matrix and was used for all further experiments. Three different Dowex 1 resins were tested, all of mesh size 200–400 but differing in the degree of cross-linkage (2, 4, 8%). For each, the dry weight of AJ270 adsorbed increased with the dry weight of bacteria added over the range of 2–8 mg dry weight per cm^3 settled resin, but little further binding was observed on increasing the amount of bacteria added further. Dowex $1 \times 4-400$ (4% cross-linked) bound slightly more bacteria (7.6 ± 0.3 mg dry weight AJ270 per cm^3 settled resin; 15.4 mg dry weight per g dry resin) than the other two resins (7.0 ± 0.2 mg/cm^3 for 2% cross-linked and 6.4 ± 0.2 mg/ml for 8% cross-linked), and the data below were obtained with this matrix. The specific amidase activity of the bound AJ270 decreased as the bacterial loading increased so that resin loaded with 7.6 mg dry weight bacteria per cm^3 had only half the specific activity of resin carrying 2 mg dry weight per cm^3 (0.9 and 1.6 units per mg dry weight, respectively).

Under the binding conditions used, the resin reached 80% of capacity within 1 min. and full capacity after 2 min. Scanning electron microscopy of resin loaded with 7.6 mg dry weight AJ270 per cm^3 showed the adsorbed biocatalyst to consist of a monolayer of single, rod-shaped bacteria attached along their longest axis and

Table 2. Thermal stability of amidase at 60°C in AJ270 adsorbed to Dowex 1, entrapped in agar or in free suspension. Samples (1 cm^3) of AJ270 in free suspension, adsorbed to Dowex 1X4-400 at 2 mg dry weight bacteria per cm^3 resin or entrapped in agar at 20 mg dry weight per cm^3 gel were quickly heated to 60°C and held at that temperature for timed intervals. They were then rapidly cooled in an ice-water bath and assayed for amidase activity against acetamide. 100% = 3.1 units per mg dry weight for free bacteria, 1.7 units per mg dry weight for adsorbed bacteria, or 1.6 units per mg dry weight for entrapped bacteria

	% of initial amidase activity		
Time (min)	Free bacteria	Adsorbed bacteria	Entrapped bacteria
0	100	100	100
5	74	96	98
10	35	61	70
20	31	53	62
40	26	44	47
60	19	34	38

evenly covering the surface of the resin beads. This suggests that the groups involved in attachment are evenly distributed over the bacterial surface whilst cell-to-cell interactions are not important in attachment. Transmission electron microscopy of bead cross-sections confirmed that the bacteria did not penetrate to the interior of the resin beads.

Effect of salt, temperature, and pH value on the amidase activity of adsorbed AJ270

Adsorbed amidase activity was not eluted by prolonged washing of the immobilized biocatalyst in a buffer containing 2.5 *M* NaCl indicating that factors other than charge are important in attachment. The temperature profile for the amidase activity of immobilized bacteria (tested at 0.2 cm^3 resin or 0.4 mg dry weight AJ270 per cm^3 buffer) was very similar to that of bacteria in free suspension (0.4 mg dry weight per cm^3 buffer), activity increasing steadily to a maximum at about 55°C with only 60% remaining at 70°C. The amidase activity of the adsorbed bacteria was significantly more stable to high temperature than that of bacteria in free suspension (Table 2) with a significant three-fold increase in $T_{0.5}$ from 7 min to 22 min. The *pH* profiles for adsorbed and free bacteria were very similar at high *pH* values with clear optima at $pH = 8.0$, but the adsorbed bacteria were significantly less active at acid *pH* values (Fig. 1). This effect was not due to elution of adsorbed bacteria with decreasing *pH* since washing the immobilized biocatalyst with buffers of *pH* value from 5.5 to 9.0 and then measuring their amidase activity at *pH* 8.0 resulted in losses of amidase activity not exceeding 4%.

Entrapment of AJ270 in agar and agarose beads

There are several gels that may be used for the entrapment of bacteria differing in their properties and method of preparation. In this study, agar was chosen for its

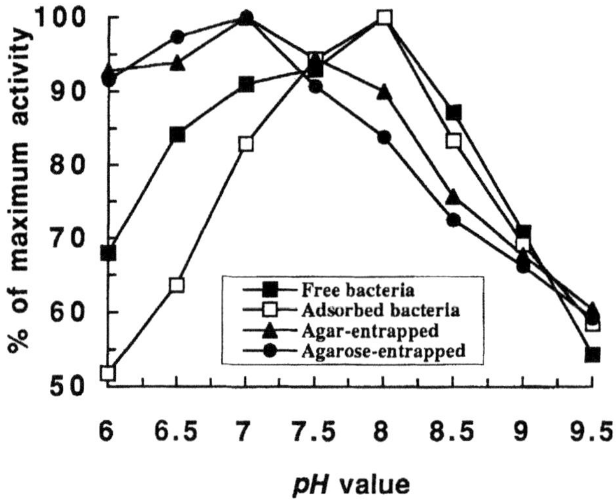

Fig. 1. Effect of *pH* on the amidase activity of *Rhodococcus* AJ270. AJ270 adsorbed to Dowex 1X4-400 beads at 2 mg dry weight per cm^3 of resin entrapped in agar or agarose beads at 20 mg dry weight per cm^3 gel or free in suspension were assayed for amidase activity against acetamide in buffers of different *pH* value. 100% = 3.3 units per mg dry weight for free bacteria, 1.7 units per mg dry weight for adsorbed bacteria, or 1.6 units per mg dry weight for entrapped bacteria

relatively low cost and ease of handling, and compared with agarose, its more expensive counterpart. When bacteria are entrapped within a carrier gel, the rate of reaction is dependent not only on the specific activity of the bacteria but also upon mass transfer limitations imposed by the topography of the carrier matrix. If the mass of bacteria within the gel is increased, there will be a point at which the rate of reaction becomes diffusion limited, and further increases in bacterial loading will have little effect. This point occurred above 20 mg dry weight bacteria (per cm^3 gel), equivalent to 1.33 g dry weight (per g dry matrix) with agar and agarose gels, under the conditions used in this study (1.5% w/v gels; bead size < 1 mm diameter; 50 m*M* acetamide; assay conditions as described in the Experimental section) (Fig. 2). Bacterial loadings in excess of 80 mg dry weight (per cm^3 gel) were readily achieved with agar and agarose gels but with no further increase in amidase activity per g gel.

Effect of temperature and pH on the amidase activity of entrapped AJ270

The temperature-activity profiles for the amidase activity of bacteria entrapped in agar or agarose (tested at 0.2 cm^3 gel or 4 mg dry weight AJ270 per cm^3 buffer) were almost identical to those reported above for bacteria in free suspension or for bacteria adsorbed to Dowex 1. Amidase activity was most stable in agar-entrapped bacteria with $T_{0.5}$ at 60°C = 34 min, a more than four-fold increase over the $T_{0.5}$ of free bacteria (Table 2). The *pH* optimum of amidase activity was shifted to lower *pH* values in both agar- and agarose-entrapped bacteria (Fig. 1) with highest activities observed at *pH* values between 6 and 7, whereas the activity of bacteria in free suspension was highest at *pH* 8.0 and was 32% lower at *pH* 6.

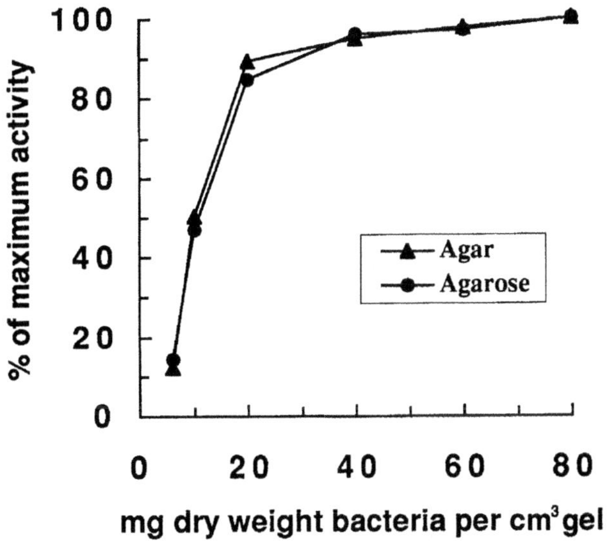

Fig. 2. Effect of bacterial loading on the specific amidase activity of *Rhodococcus* AJ270 entrapped in agar beads; 100% = 32 units per cm^3 gel)

Kinetics of amidase activity in free and immobilized AJ270

Apparent K_m and V_{max} values for acetamide and acrylamide hydrolysis by AJ270 washed suspensions and by Dowex-adsorbed, agar- and agarose-entrapped AJ270 are given in Table 3. The K_m values for free bacteria and for adsorbed bacteria were

Table 3. Kinetic data for amidase activity in free and immobilized AJ270. K_m and V_{max} values were determined by *Lineweaver-Burke* double reciprocal plots of specific amidase activity as a function of substrate concentration; n.a.: not applicable, n.d.: not done

Parameter	Immobilisation Matrix			
	None	Dowex 1	Agar	Agarose
Loading (mg dry weight per cm^3)	n.a.	20	20	20
Apparent K_m for acetamide (m*M*)	6.7±0.4	7.3±0.5	20.4±0.8	20.1±0.9
Apparent V_{max} for acetamide (*U* per mg dry wt. bacteria)	3.9±0.3	2.4±0.3	2.2±0.2	2.1±0.3
Apparent V_{max} for acetamide (*U* per cm^3 matrix)	n.a.	5	48	42
Apparent K_m for acrylamide (m*M*)	9.2±0.3	11.1±0.5	24.7±0.9	n.d.
Apparent V_{max} for acrylamide (*U* per mg dry wt. bacteria)	1.1±0.2	0.8±0.2	0.7±0.2	n.d.
Apparent V_{max} for acrylamide (*U* per cm^3 matrix)	n.a.	1.6	14	n.d.

very similar, presumably because the monolayer of bacteria adsorbed to the beads of anionic exchanger does not significantly affect the free diffusion of substrate to the bacteria. In contrast, bacteria entrapped in gel beads are impeded in their access to substrate, and this is reflected in the three-fold higher apparent K_m value. Nevertheless, because of the higher bacterial loadings achieved by entrapment, this process resulted in an immobilized biocatalyst with at least 8-times the activity (per unit volume) achieved with Dowex-adsorbed bacteria (based on V_{max} values).

We conclude that adsorption has little effect on amidase kinetic parameters, but significantly improves the temperature stability. As might be expected, entrapment in agar or agarose beads has a greater effect on kinetic parameters producing a biocatalyst with higher K_m. However, this disadvantage is outweighed by (*i*) much higher bacterial loadings, (*ii*) greater temperature stability, and (*iii*) a shift in the *pH* optimum from 8 to 7. As pointed out by *Brennan et al.* [5], *pH* 7 is ideal for ammonium acrylate production. In addition, the physical characteristics of the gel-entrapped biocatalysts are better suited to continuous column reactors than Dowex 1, allowing for higher flow-rates at lower pressure differentials. No advantage was found in using the more expensive agarose, rather than agar, as the immobilization matrix.

Biotransformation of acrylamide to acrylic acid in an STR

The kinetic data for acrylamide hydrolysis by agar-entrapped AJ270 (Table 3), in comparison with the data for acetamide hydrolysis, indicated a somewhat poorer affinity for this substrate and a significantly lower reaction rate. Nevertheless, the data suggest that entrapped AJ270 could be a very useful biocatalyst for acrylic acid production. In order to test this, the hydrolysis of 250 m*M* acrylamide to acrylic acid was attempted using 10 cm^3 of agar-entrapped bacteria (20 mg dry weight bacteria per cm^3) in a 50 cm^3 batch STR at 30°C (Fig. 3). Under the conditions used (see Experimental), acrylamidase activity was independent of stirrer speeds between 100 and 500 rpm. A STR operated in batch mode may be subject to diffusional limitations represented by the stationary boundary layer between the bulk fluid and the support surface. Clearly even the lowest stirrer speed used in this study was sufficient to overcome this boundary layer effect. About one fifth (2.44 mmol) of the starting acrylamide (12.5 mmol) was converted to acrylic acid after 30 min (specific activity for acrylamide hydrolysis = 0.4 units per mg dry weight bacteria), but the reaction then slowed down very significantly so that after 210 min only 59% of the acrylamide had been hydrolyzed. The values quoted are the mean values of three separate experiments, and in each case at least 95% of the acrylamide hydrolyzed was accounted for as acrylic acid.

Inhibition of amidase activity by acrylic acid

The results in Fig. 3 suggest that contact with high concentrations of acrylamide or acrylic acid inhibits the amidase activity of the immobilized biocatalyst. Exposure to high acrylamide concentrations was reduced by stepwise addition of the substrate (Fig. 3). This increased the final concentration of acrylic acid achieved, but only slightly improved the proportion of acrylamide hydrolysed (from 59 to

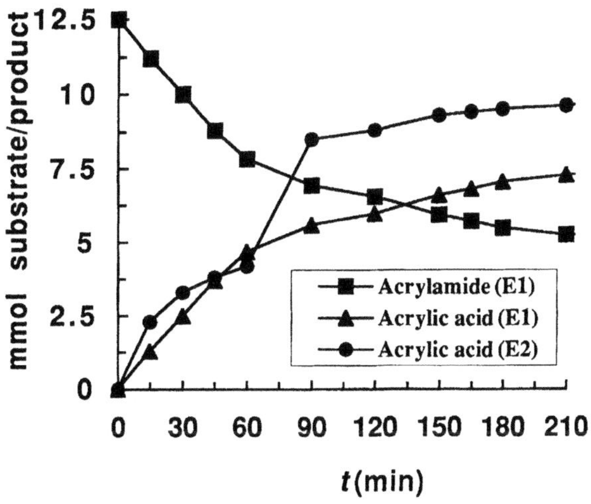

Fig. 3. Hydrolysis of acrylamide by agar-entrapped *Rhodococcus* AJ270 in a STR. In the first experiment (E1), agar-entrapped AJ270 (10 cm^3; 20 mg dry weight per cm^3 gel) was incubated with 12.5 mmol acrylamide in a 50 cm^3 STR. In the second experiment (E2), the STR contained 5 mmol acrylamide, and further 5 mmol were added at 60, 120 and 180 min. Acrylamide and acrylic acid concentrations were determined by GC

66%), suggesting that acrylic acid is the major inhibitor. This was confirmed (Table 4) by incubating entrapped bacteria for 20 min with 0–250 m*M* acrylic acid, then washing the biocatalyst with buffer, and measuring its amidase activity against acetamide. Inhibition by acrylic acid in the experiments described above was not due to a significant fall in *pH*.

Continuous acrylamide hydrolysis/acrylic acid production in a column reactor

In a continuous column reactor fed with an acrylamide solution, only the biocatalyst at the outlet end of the column is exposed to high concentrations of product and therefore subject to inhibition by acrylic acid. The amidase activity of such a column is unlikely to suffer the dramatic losses observed in the STR.

Continuous column reactors containing 40 cm^3 of agar-entrapped AJ270 were challenged with acrylamide solutions of different concentrations at various flow

Table 4. Inhibition of amidase activity in entrapped AJ270 by acrylic acid. Samples of AJ270 entrapped in agar beads at 20 mg dry weight per cm^3 gel were incubated at 4°C for 20 min in buffer containing various concentrations of acrylic acid. The beads were then spun, washed in buffer, and assayed for amidase activity against acetamide. 100% = 31 units per cm^3 gel.

Acrylic acid (m*M*)	% Activity remaining
0	100
25	86
50	81
100	74
175	61
250	50

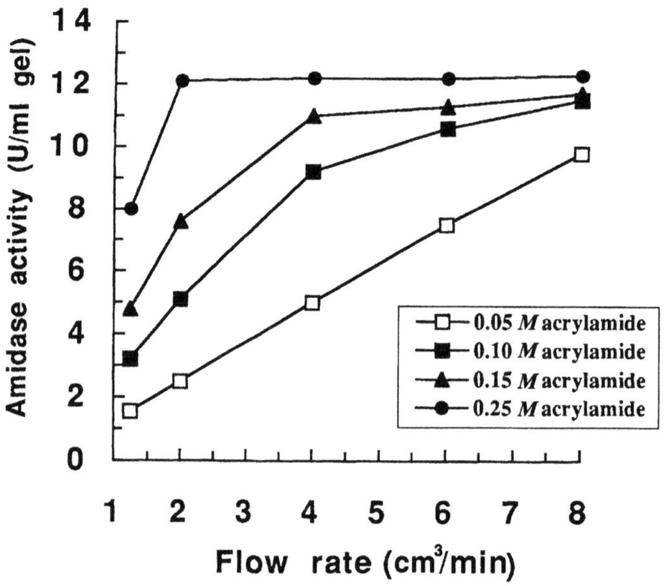

Fig. 4. Effect of flow rate and acrylamide concentration on the amidase activity of agar-entrapped *Rhodococcus* AJ270 in a column reactor. Agar-entrapped AJ270 (40 cm^3 gel; 20 mg dry weight per cm^3) in a column reactor (internal diameter 2.4 cm) was fed with buffers containing 0.05–0.25 *M* acrylamide at various flow rates. Acrylamidase activities were computed from the acrylamide concentration in the input and effluent streams

rates. Specific activity increased with both flow rate and acrylamide concentration until a plateau was reached at about 12.3 units (per cm^3 gel) (Fig. 4). As expected, % conversion rates followed the reverse trend (Fig. 5). The results suggested that a flow rate of 1.3–2.0 $cm^3 \cdot min^{-1}$ would achieve >95% conversion of 0.25 *M* acrylamide.

In a final set of experiments, similar column reactors packed with agar-entrapped AJ270 (40 cm^3 gel; 20 mg dry weight bacteria per cm^3 of gel) and maintained at 30°C were fed continuously with 0.25 *M* acrylamide in 0.1 *M tris*-HCl buffer (*pH* 7) at 1.5 $cm^3 \cdot min^{-1}$ for 30 h. These reactors were operated at >97% conversion of the substrate to acrylic acid for 22 h (specific acrylamidase activity = 0.4 units per mg dry weight bacteria), after which the % conversion fell steadily to 60% at 30 h.

Rhodococci expressing nitrile hydratase and immobilized in polyacrylamide gel are used for the commercial production of acrylamide from acrylonitrile in a plant with a capacity of 30000 t/a [7]. In contrast, a commercial process for the production of acrylate has not been developped, even though this commodity chemical is widely used in the production of polymeric flocculents, thickeners, adhesives, and adsorbents. Several workers have described the isolation of bacteria able to grow on acrylamide [8–11], but most work on acrylamide hydrolysis, as in the present study, has involved bacteria grown on simpler, saturated aliphatic nitriles and amides.

Rhodococcus AJ270 grown on acetamide expresses high amidase activity against acrylamide. However, with 0.25 *M* acrylamide as substrate, exposure to the acrylic acid product slowly inactivates the amidase activity, eventually limiting the

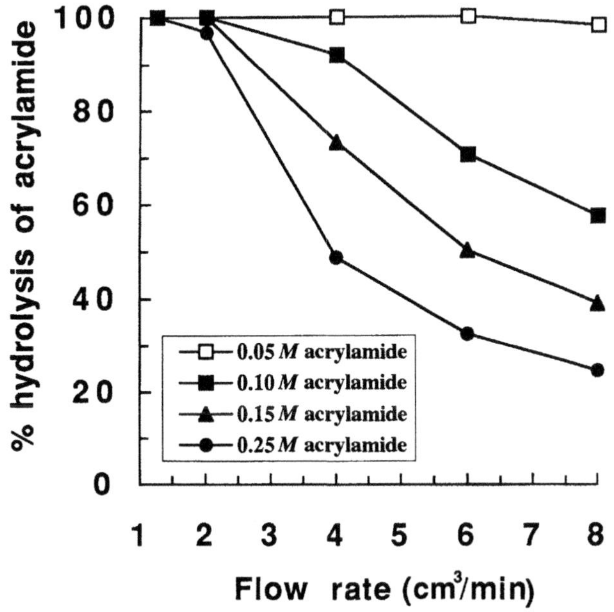

Fig. 5. Effect of flow rate and acrylamide concentration on percent conversion by agar-entrapped *Rhodococcus* AJ270 in a column reactor. The conditions were as for Fig. 4; percent conversion figures were computed from the acrylamide concentrations in the input and effluent streams

% conversion of the acrylamide substrate. Inactivation of immobilized acetamide-grown *Corynebacterium nitrilophilus* by high concentrations of acrylamide and acrylic acid during the biocatalytic hydrolysis of acrylamide has been noted previously [5] and is thought to be due to the effect of these β-unsaturated compounds on sulfydryl groups essential for amidase activity [12]. In the present study, inactivation was minimized by using a fixed-bed column reactor allowing continuous production of acrylic acid for 22 h at >97% conversion of acrylamide. Improvements on this basic protocol, such as the periodic replacement of the biocatalyst at the outlet end of the reactor, might enable extended production of acrylate using this biocatalyst.

Experimental

Chemicals and immobilization matrices

Acrylic acid (99%) and acetamide (99+%) were obtained from Aldrich (Dorset, UK). Dowex 1X4-400, purified agar, agarose (type 1-A), and all other chemicals and matrices were purchased from Sigma (Dorset, UK). Vegetable oil (pure sunflower oil) was obtained from a large UK retail outlet (ASDA Stores, Leeds). The ion exchange resins used in the study (see Table 5) were all polystyrene based. Amberlite IRA900 and IRA904 are macroreticular, the others are gels.

Analytical methods

Acrylamide and acrylic acid were determined by gas chromatography as described previously [2]. The concentrations of ammonia in assay samples were determined using *Nessler*'s reagent. The

Table 5. Properties of the anionic exchange resins used

Resin	Active group	Mesh size	% Cross linked	Capacity ($meq \cdot cm^{-3}$)	Ionic form
Amberlite IRA904	quaternary ammonium	20–50	–	1.0	Cl^-
Amberlite IRA400	quaternary ammonium	16–50	8	1.4	Cl^-
Amberlite IRA402	quaternary ammonium	16–50	6	1.3	Cl^-
Amberlite IRA440c	quaternary ammonium	16–40	–	1.0	OH^-
Amberlite IRA900	quaternary ammonium	16–50	–	1.0	Cl^-
Dowex 1X8-400	trimethylbenzylammonium	200–400	8	1.2	Cl^-
Dowex 2X8-400	dimethylethanol-benzylammonium	200–400	8	1.2	Cl^-

sample ($0.025\,cm^3$) was diluted with $0.85\,cm^3$ of water, and then $0.125\,cm^3$ of reagent (ammonia colour reagent from Sigma, Dorset, UK) were added. The colour was allowed to develop for 10 min before measuring the absorbance at 420 nm. Standard curves were prepared using NH_4Cl (>99.5%). Bacterial concentrations of washed suspensions were determined turbidometrically by measuring their absorbance at 540 nm and converted to dry weights using a standard curve. The data are the mean values from three separate experiments; in no case the results of single experiments differed by more than 12%.

Bulk growth of Rhodococcus AJ270

AJ270 was grown on 50 m*M* acetamide, and washed suspensions were prepared as described previously [1].

Adsorption on Dowex 1 and other anionic exchangers

Bacteria and resin were both washed three times in 0.1 M *tris*-HCl buffer, *pH* 8.0, prior to use. Bacteria were re-suspended in the same buffer to 22 mg dry weight (per cm^3) and $5\,cm^3$ of the suspension were added to a Universal bottle containing 3 g of the washed resin. The mixture was agitated for 45 min on a Denley Spiramix, the resin allowed to settle for 5 min, and then the supernatant was removed. The settled resin was washed five times with 0.01 *M tris*-HCl buffer, *pH* 8.0, to remove any unadsorbed bacteria. The amount of bacteria adsorbed was determined by comparing the absorbance (A_{540nm}) of the bacterial suspension before and after adsorption. The basic method was modified by changing the ratio of bacteria to resin, including *EDTA* in the washing buffers, or by using different anionic exchangers.

Entrapment in agar or agarose

The method was modified from *Wikstrom et al.* [13]. Bacteria were washed three times with 0.1 *M tris*-HCl buffer, *pH* 8.0, resuspended in the same buffer to the desired concentration, and mixed with an equal volume of 3% (w/v) molten agar or agarose at 48°C. The mixture was quickly added to an equal volume of vegetable oil also held at 48°C and agitated with a *Rushton*-style impeller at 250 rpm to form an even emulsion. The temperature of the emulsion was then lowered by immersion of the vessel in an ice-water bath whilst maintaining the agitation. Once the temperature had fallen below 20°C, agitation was stopped, and the two phases were allowed to separate. The aqueous phase consisted of a heterogeneous mix of bead sizes < 1 mm in diameter. The beads were washed with buffer until no oil was visible on the buffer surface.

Bioreactors

The stirred-tank reactor (STR) consisted of a glass vessel of working volume 50 cm^3, fitted with baffles and agitated using a *Rushton*-style impeller at 100–500 rpm. Fixed-bed column reactors consisted of jacketed glass columns (2.4 cm internal diameter) packed with 40 cm^3 of agar-entrapped biocatalyst. Substrate solution was pumped through the columns in an upward direction to prevent compression of the gel. Both types of reactors were kept at constant temperature (30°C) by connecting their water jackets to a circulating water bath.

Determination of amidase activity

For washed suspensions and Dowex 1-immobilized bacteria, samples (0.1 cm^3) and 0.35 cm^3 of 0.1 M *tris*-HCl buffer, *pH* 8.0 were placed in an *Eppendorf* tube and pre-warmed to 30°C in a water bath. Substrate (0.05 cm^3 of 0.5 *M* sodium acetamide unless stated otherwise) was added, and the tube was incubated at 30°C for 5–30 min before stopping the reaction with 0.025 cm^3 of 0.25 M HCl. The mixture was then centrifuged for 2 min in a microcentrifuge, and the clear supernatant was removed for ammonia determination.

Gel-entrapped bacteria were drained of all buffer and 5 g removed to a 50 cm^3 conical flask. The beads were resuspended in 22.5 cm^3 of 0.1 *M tris*-HCl buffer, *pH* 8.0, and warmed to 30°C in a shaking waterbath. Pre-warmed substrate solution (2.5 cm^3 of 0.5 *M* sodium acetamide unless otherwise stated) was added, and samples were removed at timed intervals for ammonia determination.

In experiments where amidase activity was determined at different *pH* values, 0.1 *M* sodium phosphate buffers were used for *pH* 6–7.0 and 0.1 *M tris*-HCl buffers for *pH* 7.5–9.5.

Electron microscopy of adsorbed bacteria

Samples were fixed in 2.5% (w/v) glutaraldehyde in 0.1 M sodium phosphate buffer, *pH* 7.3, at 4°C. The fixed samples were then washed twice with the same buffer before fixing in 1% (w/v) osmium tetroxide for 1 h. After washing twice with water, the samples were fixed in 1% (w/v) uranyl acetate for 30 min and washed in water for 5 min. After fixing, samples were dehydrated by washing with ethanol (50, 70, 90% v/v and then twice with 100% for 15 min each wash).

Samples for TEM were embedded in epoxy resin, sectioned, and the sections mounted on copper grids. Grids were stained with saturated uranyl acetate in 70% (v/v) ethanol for 10 min, washed with water, then with alkaline lead citrate for 10 min, and then with water again before drying. Samples for SEM were mounted on carbon adhesive discs and sputter-coated initially with carbon and then gold/palladium. Specimens were viewed using a Hitachi H7000 microscope and, for SEM, a H7110 scanning attachment.

Acknowledgements

This work was supported in part by research grant GR/J58299 from BBSRC.

References

[1] Snell D, Colby J (1999) Enz Microbial Technol **24**: 160
[2] Blakey AJ, Colby J, Williams E, O'Reilly C (1995) FEMS Microbiol Lett **129**: 57
[3] Meth-Cohn O, Wang M-X (1997) J Chem Soc Perkin Trans 1099
[4] Meth-Cohn O, Wang M-X (1997) J Chem Soc Chem Commun 1041
[5] Brennan MR, Armitage YC, Mortimer MG, Hughes J, Ramsden DK (1995) Biotechnol Lett **17**: 513
[6] Nagasawa T, Yamada H (1989) Trends Biotechnol **7**: 153

[7] Yamada H, Kobayashi M (1996) Bioscience Biotechnol Biochem **60**: 1391
[8] Nawaz MS, Franklin W, Cerniglia CE (1993) Can J Microbiol **39**: 207
[9] Nawaz MS, Franklin W, Cerniglia CE (1994) Environ Sci Technol **28**: 1106
[10] Kumar A, Kumar A (1998) J Microbiol Biotechnol **8**: 347
[11] Nawaz MS, Khan AA, Seng JE, Leakey JE, Siitonen PH, Cerniglia CE (1994) Appl Environ Microbiol **60**: 3343
[12] Cavins JF, Friedman M (1968) J Biol Chem **143**: 3357
[13] Wikstrom P, Szwajcer E, Brodelius P, Nilsson K, Mossbach K (1982) Biotechnol Lett **4**: 153

Received November 18, 1999. Accepted December 14, 1999

Beauveria bassiana ATCC 7159 Contains an *L*-Specific α-Amino Acid Benzamidase

Herbert L. Holland*, **Peter R. Andreana**, **Reza Salehzadeh-Asl**, **Aaron van Vliet**, **Nancy J. Ihasz**, and **Frances M. Brown**

Institute of Molecular Catalysis, Department of Chemistry, Brock University, St. Catharines, ON L2S 3A1, Canada

Summary. Biotransformation of a series of racemic N-benzoyl α-amino acids by the fungus *Beauveria bassiana* ATCC 7159 results in isolation of the corresponding *D*-amino acid benzamides in high enantiomeric purity and yield.

Keywords. Amino acid; *Beauveria bassiana*; Biocatalysis; Biotransformation, Kinetic resolution

Introduction

Enzymatic methods for the kinetic resolution of racemic α-amino acids are well established [1]. Methods have been developed that involve the stereoselective hydrolysis of carboxylic acid ester [2] or amino amide [3] derivatives; either *D*- or *L*-specific hydrolysis can occur, and the product may be isolated as the free amino acid or as the unhydrolyzed substrate [4]. These transformations are summarized in Fig. 1.

Enantioselective hydrolyses involving amide derivatives of the amino group are most frequently reported using an amino acylase enzyme preparation [5] or a whole cell biocatalyst containing such an enzyme [6] for hydrolysis of the acetamide group. Reports of the kinetic resolution of other amino amide derivatives of α-amino acids by aminoacylase enzymes include *inter alia* examples of hydrolyses of N-chloroacetyl [7], N-trifluoroacetyl [7], N-methoxyacetyl [7], N-benzyl [8], and N-benzyloxycarbonyl [8] derivatives.

The enzymatic hydrolysis of N-benzoyl (benzamide) derivatives of α-amino acids has been only rarely reported, in spite of an early account of the use of takadiastase for the enantioselective hydrolysis of such compounds [9]. A "*D*-deacylase powder" from an unidentified pseudomonad has been reported to hydrolyze the *D* enantiomer of N-benzoylphenylalanine in an enantioselective manner [10], and *L*-N-benzoylglutamic acid is converted to the free amino acid by an enzyme described as urethane hydrolase IV from an unidentified microbial source [11], but the hydrolysis of N-benzoyl derivatives has not been systematically

* Corresponding author

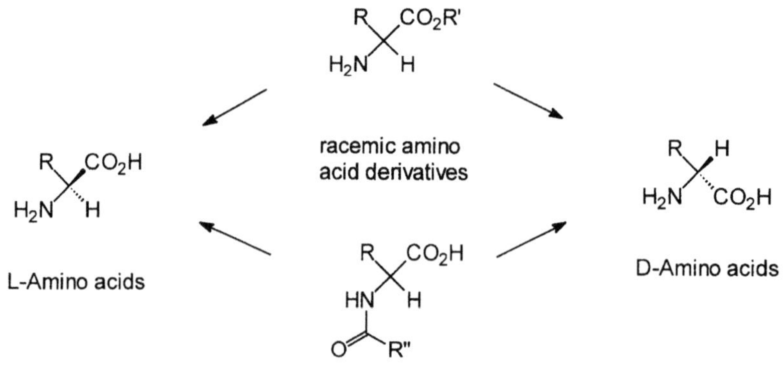

Fig. 1. Kinetic resolution of amino acid derivatives

investigated as a means for kinetic resolution of the naturally occurring α-amino acids.

The fungus *Beauveria bassiana* ATCC 7159 is one of the most frequently used whole cell biocatalysts, being exceeded in application only by baker's yeast, *Pseudomonas putida*, and *Aspergillus niger* [12]. It catalyzes a wide variety of reactions which have recently been reviewed [13], including the hydrolysis of amides. In this report we describe the use of *B. bassiana* for the *L*-specific enantio-selective hydrolysis of a series of N-benzoyl α-amino acids.

Results and Discussion

Biotransformation of the N-benzoyl derivatives of racemic α-amino acids by *Beauveria bassiana* ATCC 7159 results in the enantioselective hydrolysis of the natural *L*-isomer, leaving the residual *D*-N-benzoyl amino acid in high enantiomeric excess. These results are summarized in Fig. 2 and Table 1.

The enantiomeric excess (*ee*) of the products listed in Table 2 were determined by chiral stationary phase HPLC analysis with reference to authentic enantiopure standards. Yields refer to isolated, purified material and are based on the total amount of racemic substrate used. Low yields in the case of the sulfur-containing amino acids may be attributable to the formation of the corresponding sulfoxides as side products. *Beauveria bassiana* is known to convert N-protected thia-amino acids to the corresponding sulfoxides with high diastereomeric selectivity [14], but this reaction is slower than hydrolysis of the benzamide group, taking 3–5 days for completion compared with the 1–2 days required for the hydrolytic process. A time course study for the hydrolysis of *DL*-N-benzoylalanine is presented in Fig. 3, from which it is apparent that complete conversion of the *L*-enantiomer has been

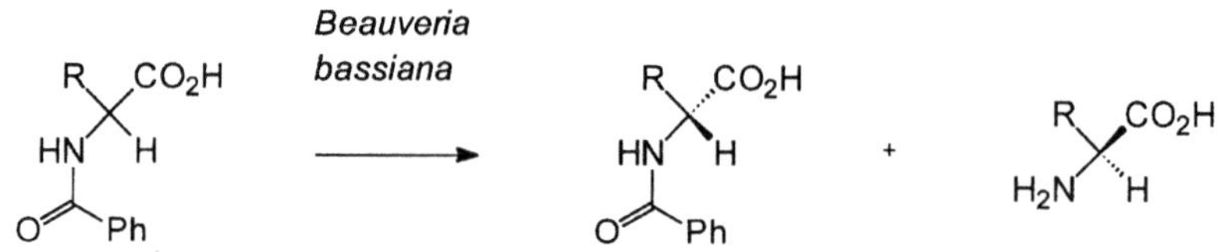

Fig. 2. Kinetic resolution of N-benzoyl amino acid derivatives by *B. bassiana* ATCC 7159

Table 1. Kinetic resolution of N-benzoyl amino acid derivatives by *B. bassiana* ATCC 7159

(±)-N-Benzoyl α-amino acid substrate		*D*-N-Benzoyl α-amino acid product	
	Substituent	Yield/%	*ee*/%
Alanine	CH_3	20	>95
Asparagine	CH_2CONH_2	28	80
Aspartic acid	CH_2CO_2H	23	>95
Cysteine	CH_2SH	17	>95
Ethionine	$CH_2CH_2SCH_2CH_3$	16	>95
Glutamic acid	$CH_2CH_2CO_2H$	22	>95
Glutamine	$CH_2CH_2CONH_2$	25	20
Leucine	$CH_2CH(CH_3)_2$	23	>95
Methionine	$CH_2CH_2SCH_3$	18	>95
Phenylalanine	$CH_2C_6H_5$	30	>95
Phenylglycine	C_6H_5	27	>95
Pipecolinic acid	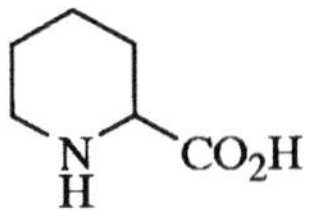	20	>95
Proline	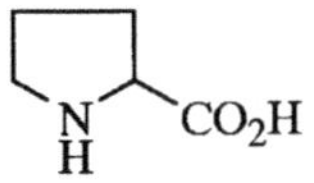	20	>95
S-Methylcysteine	CH_2SCH_3	32	>95
Tryptophan	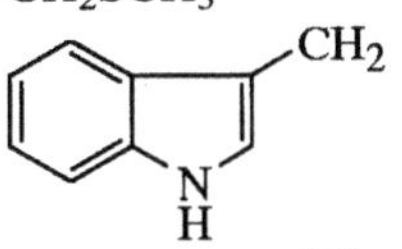	21	>95
Tyrosine	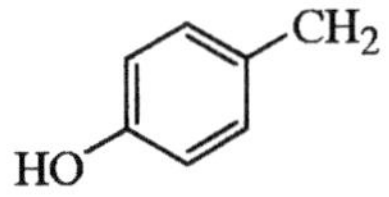	17	20
Valine	$(CH_3)_2CH$	28	>95

[a] Refer to structures of Fig. 2; complete structures are shown for pipecolinic acid and proline only

attained after 36–48 hours incubation. Similar studies were not performed for the other substrates listed in Table 2, but on the basis of the data presented in Table 2 a 72 hour incubation period was selected as standard. The moderate *ee* observed for hydrolyses of racemic N-benzoyl asparagine, glutamine and tyrosine may be attributable to incomplete conversion of these substrates over the period of their biotransformation.

The hydrolysis of benzamides by *B. bassiana* appears to be selective for α-amino acid derivatives. Benzamides of β-amino acids such as 3-aminobutyric acid and 3-amino-3-phenylpropionic acid were not hydrolyzed by *B. bassiana*, substrates being recovered unchanged following incubation for five days. In other studies [15–19] many different amino substrates protected as their N-benzoyl derivatives have been hydroxylated by *B. bassiana* without loss of the benzoyl group; indeed, presence of the latter functionality in the substrate is often beneficial to the yield and stereoselectivity of such processes, as the benzamide unit may act as a binding group for the hydroxylase enzymes present in *B. bassiana* [18, 19]. As shown in the present

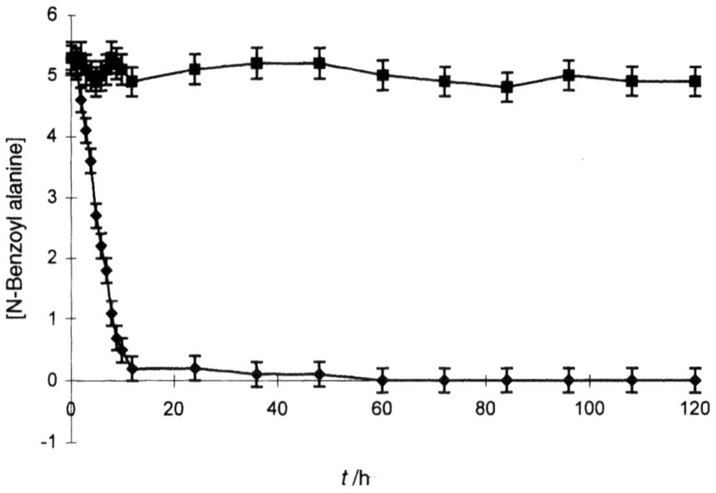

Fig. 3. Time course for the hydrolysis of *DL*-N-benzoylalanine by *B. bassiana* ◆: *L*-N-benzoyl alanine, ■: *D*-N-benzoyl alanine

study, however, the hydrolysis of racemic N-benzoyl α-amino acids by *B. bassiana* presents a simple method for the preparation of the *D*-benzoyl α-amino acids listed in Table 1 in moderate yield and, for most examples, high enantiomeric purity.

Experimental

Materials and methods

Melting points were determined on a Kofler hot stage and are uncorrected. The ^{1}H NMR spectra were recorded on a Bruker Avance series 300 spectrometer in CD_3OD solution using residual CH_3OH as the internal standard; chemical shifts are reported in ppm (δ), and the multiplicities are quoted as s (singlet), d (doublet), t (triplet), q (quartet), or m (multiplet). The ^{13}C NMR spectra were recorded at 75 MHz on the same spectrometer in CD_3OD solution. Mass spectra were obtained in the +FAB mode using a Kratos 1S spectrometer. IR spectra were obtained using a Mattson Research Series FTIR spectrometer. Optical rotations were recorded at ambient temperature in methanol as solvent using a Rudolph Autopol 3 polarimeter. Enantiomeric excess (*ee*) was determined using an ASTEC Chirobiotic T^{TM} column 250×4.6 mm with acetonitrile/methanol/acetic acid/triethylamine 455/545/2/2 solvent at a flow rate of 1 cm^3/min with UV detection at 254 nm. TLC was performed on Merck silica gel F_{254} plates (0.2 mm), and column chromatography used Merck silica gel 9385 (230–400 mesh). *Beauveria bassiana* ATCC 7159 was maintained on *Sabouraud* dextrose agar slopes, grown at 26°C and stored at 4°C.

Preparation of substrates

With the exception of the N-benzoyl tyrosine, all substrates were prepared in 85–95% yield by the following general procedure: the racemic amino acid (1 g) was dissolved in sodium hydroxide solution (20 cm^3, 2 *M*), and the solution cooled to 0°C. Benzoyl chloride (1.1 mole equivalents) was added in ten portions over a period of 1 h while the solution was stirred vigorously. The *pH* was checked periodically and adjusted if necessary to >12 by the addition of 2 *M* NaOH. When the addition was complete, the solution was stirred at room temperature for 2 h, then re-cooled to 0°C and adjusted to $pH=2$ by the addition of 5% HCl. The precipitated product was separated by filtration, dried, and crystallized from 95% ethanol. Products from proline and pipecolinic acid did not precipitate in crystalline form and were isolated by extraction with CH_2Cl_2 and evaporation of the extract to give a solid residue which was then crystallized. All products gave satisfactory spectroscopic data and melting points in agreement with those reported [20].

N-Benzoyltyrosine was prepared by a modification of the above procedure in which methanol (10 cm^3) was added to the reaction mixture following the 2 h stirring period at room temperature. The mixture was then stirred at room temperature overnight and subsequent steps performed as described above.

Biotransformations with B. bassiana

A growth medium (1 dm^3) composed of glucose (10 g) and corn steep liquor (20 g) in distilled water, adjusted to *pH* 4.85 with 1 *M* NaOH, was distributed in 5 1 dm^3 *Erlenmeyer* flasks which were stoppered with foam plugs and sterilized by autoclaving at 121°C for 20 min. The flasks were allowed to cool and then inoculated under sterile conditions with *B. bassiana* taken from a 3-day old agar slope. The flasks were allowed to stand overnight at 27°C and then placed on a rotary shaker (1″orbit) at 180 rpm, 27°C. After three days, a solution of the appropriate substrate (0.5 g) in 95% ethanol (10 cm^3) was added, and growth was allowed to continue for a further 3 days. The fungal mass was removed by filtration, and the filtrate was adjusted to *pH* 3 and continuously extracted with CH_2Cl_2 for 4 days. The extract was then evaporated to give a residue that was examined by TLC and chiral stationary phase HPLC and subjected to column chromatography using ethyl acetate/methanol solvent mixtures containing 2% acetic acid. The solvent composition was changed from 0% methanol to 40% methanol in 10% steps, and elution with the final mixture was continued until all product was removed from the column. All products had physical constants (melting points and optical rotation data) as reported [20] and exhibited spectroscopic data consistent with their structures.

Acknowledgements

This work was supported by the Natural Sciences and Engineering Research Council of Canada

References

[1] Faber K (1997) Biotransformations in Organic Chemistry, 3rd edn. Springer, Berlin, p 50
[2] Miyazawa T (1999) Amino Acids **16**: 191
[3] Drauz, K, Waldmann H (eds) (1995) Enzyme Catalysis in Organic Synthesis VCH, Weinheim, p 379
[4] Kamphuis J, Boesten, WHJ, Kaptein B, Hermes HFM, Sonke T, Broxterman QB, van den Tweel WJJ, Schoemaker HE (1992) The Production and Uses of Optically Pure Natural and Unnatural Amino Acids. In: Collins AN, Sheldrake GN, Crosby J (eds) Chirality in Industry. Wiley, Chichester, p 187
[5] Wong CH, Whitesides GM (1994) Enzymes in Synthetic Organic Chemistry. Pergamon, Oxford, p 42
[6] Glänzer BI, Faber K, Griengl H (1987) Tetrahedron **43**: 771
[7] Chenault HK, Dahmer J, Whitesides GM (1989) J Am Chem Soc **111**: 6354
[8] Yang YB, Lin CS, Tseng CP, Wang YJ, Tsai YC (1991) Appl Environ Microbiol **57**: 1259
[9] Neuberg C, Mandl I (1950) Enzymologia **14**: 128
[10] Kameda Y, Toyoura E, Kimura Y (1958) Nature **181**: 1225
[11] Matsumura S, Shin T, Murao S, Sakaguchi M, Kawano T (1985) Agric Biol Chem **49**: 3643
[12] Biocatalysis Database (1999) Synopsys Scientific, Leeds, UK, http://www.synopsys.co.uk

[13] Grogan GJ, Holland HL (1999) J Mol Catal (B) (in press)
[14] Holland HL, Andreana PR, Brown FM (1999) Tetrahedron Asymm **10**: 2833
[15] Johnson RA, Murray HC, Reineke LM (1971) J Am Chem Soc **93**: 4872
[16] Herr ME, Murray HC, Fonken GS (1971) J Med Chem **14**: 842
[17] Johnson RA, Murray HC, Reineke LM, Fonken GS (1968) J Org Chem **33**: 3207
[18] Archelas A, Fourneron JD, Vigne B, Furstoss R (1986) Tetrahedron **42**: 3863
[19] Braunegg G, de Raadt A, Feichtenhofer S, Griengl H, Kopper I, Lehmann A, Weber, H-J (1999) Angew Chem Int Ed Engl **38**: 2763
[20] Buckingham J (ed) (1982) Dictionary of Organic Compounds, 5th edn. Chapman & Hall, New York

Received December 20, 1999. Accepted January 21, 2000

Bacterial Biotransformation of Isoprene and Related Dienes

Derek R. Boyd,[1,*], **David Clarke**[1], **Marcel C. Cleij**[1], **John T. G Hamilton**[2], and **Gary N. Sheldrake**[1]

[1] School of Chemistry and QUESTOR Centre, The Queen's University of Belfast, Belfast BT9 5AG, UK

[2] Food Science Division, Department of Agriculture for Northern Ireland, Belfast BT9 5PX, UK

Summary. The bacterium *Pseudomonas putida* ML 2 was used in the oxidative biodegradation of the acyclic dienes isoprene, *trans*-piperylene, *cis*-piperylene, and 1,3-butadiene. Regioselective dioxygenase-catalyzed dihydroxylation of alkenes yielded vicinal diols in the preferred sequence monosubstituted > *cis*-disubstituted > *gem*-disubstituted > *trans*-disubstituted. The isolated diol metabolites had an excess of the *R* configuration (9–97% *ee*), and further diol oxidation was controlled by addition of propylene glycol as an inhibitor. Stereoselectivity using the ML2 strain resulted from both enzymatic asymmetric alkene dihydroxylation and kinetic resolution of diols. Enantioselective oxidation of the allylic secondary alcohol group of *R* configuration yielded the corresponding unsaturated ketoalcohol; the residual diol was recovered with a large excess ($\geq$ 93% *ee*) of the *S* configuration. In addition to the enzymatic diene oxidation steps yielding unsaturated diols and ketoalcohols, evidence was also found of enzymatic alkene hydrogenation to yield saturated ketoalcohols and diols.

Keywords. Asymmetric dihydroxylation; Chiral resolution; Dioxygenases; Enzymes; Isoprene.

Introduction

Methane and isoprene **1** are reported to be present in considerable quantities in the earth's atmosphere and are the most abundant volatile hydrocarbons in the environment [1,2]. Biogenic isoprene is produced in large quantities from plants (5×10^{14} g year^{-1}), particularly in the tropical rain forests and from the oceans [1–3]. It has been estimated that isoprene is responsible for *ca.* 30% of the total non-methane organic compounds (NMOCs) emitted into the environment in North America each year [1]. Isoprene can readily be involved in radical reactions under the influence of UV radiation and will strongly influence atmospheric chemistry. The isoprene radicals produced in the earth's atmosphere can thus undergo photochemical reactions with nitrogen oxides (NO, NO_2) from combustion sources resulting in the formation of ozone in the lowest layer of the atmosphere. Whereas the fate of isoprene in atmospheric chemistry has been studied extensively [4–6],

* Corresponding author

its effect on living systems appears to have been mainly confined to animals [7–10]. Isoprene is also produced in large quantities by the petrochemical industry as a monomer for *cis*-1,4-polyisoprene elastomer, and in this context the potential risk of human exposure to higher concentrations of anthropogenic isoprene than may normally occur in nature through biogenic isoprene has initiated toxicological studies [7–10]. These animal studies have shown that the first step in the metabolic pathway of isoprene degradation involves the monooxygenase-catalyzed formation of mutagenic epoxides. Liver microsomal cytochrome P-450 enzymes and dioxygen were found to be responsible for both *mono*- and *bis*-epoxidation of **1**. Hydrolysis of the isoprene monoepoxide metabolites gave the corresponding diols **2** and **3**. Apart from these earlier reports on the metabolism of **1** in eucaryotic cells [7–9], to our knowledge the metabolic fate of isoprene or related butadienes in procaryotic (bacterial) cells has not been reported[1]. Since both plants, a major biogenic source of isoprene, and bacteria, a major source of hydrocarbon-degrading enzymes, are found to co-exist within soil, metabolism by soil bacteria could play an important role as a biological sink controlling the concentration of biogenic isoprene released into the environment. This has indeed recently been established for the first time using both *in vitro* and *in vivo* studies of the microbial consumption of atmospheric isoprene by aerobic soil bacteria in a temperate forest soil [2]. However, this microbial consumption programme did not address the question of biodegradation pathways or the nature of metabolic intermediates and is thus complementary to the results obtained in the present study.

Dioxygenase-catalyzed *cis*-dihydroxylation of mono- and polycyclic aromatic compounds using *Pseudomonas putida* UV4 has been carried out in these laboratories [11–15]. Dioxygenase enzymes are also known to be present in two wild-type strains of *P. putida*, *i.e.* ML 2, a source of benzene dioxygenase (BDO), and NCIMB 8859, a source of naphthalene dioxygenase (NDO). UV4, a mutant strain of *P. putida* contains a dioxygenase (toluene diooxygenase, TDO), but is deficient in the *cis*-diol dehydrogenase enzyme which is responsible for the conversion of arene *cis*-dihydrodiols to catechols in the corresponding wild-type strain. Non-aromatic conjugated cyclic diene and triene substrates of different ring sizes have been reported to undergo stereoselective alkene *cis*-dihydroxylation in the presence of *P. putida* UV4 [15]. It has also been demonstrated that a cyclic aliphatic alkene can be dihydroxylated by toluene dioxygenase from *P. putida* F1 [17]. Dioxygenase-catalyzed asymmetric dihydroxylation of a series of volatile acyclic conjugated dienes, elucidation of the structure and stereochemistry of the resultant diols, and the metabolic sequence involved in the formation of other derived metabolites formed from this class of NMOCs are presented and discussed here.

Results and Discussion

A major objective of this study was to elucidate the sequence of steps involved in the metabolism of dienes **1**, **7**, **15**, and **19** by different strains of the soil bacterium *P. putida* (Schemes 1–4) and to characterize and, where possible, to assign

---

[1] See note added in proof

stereochemistry to the metabolites. Due to the volatile nature of dienes **1** (b.p.: 34°C), **7** (b.p.: 42°C), **15** (b.p.: 44°C), and **19** (b.p.: –5°C) and their limited solubility in the aqueous culture medium, the isolated yields were expected to be quite low using our reported standard biotransformation conditions [11, 14, 16]. However, by modification of these biotransformation procedures and utilization of more sensitive assay methods, it was possible to identify and stereochemically assign the bioproducts. This was achieved by using a lower incubation temperature (4°C) and higher cell densities in shake flasks (optical density values of *ca.* 10 at 600 nm), by addition of a competitive inhibitor of the diol dehydrogenase (propylene glycol) during biotransformations, and by application of chiral stationary phase GC/FID, GC/MS, and GC/IR methods for analysis of culture supernatants. Despite these modifications the isolated yields of bioproducts were relatively low (< 10%) and thus currently of limited synthetic value.

Based on reported results using *P. putida* UV4 and a series of cyclic diene substrates where the corresponding *vic*-diols were obtained [15], diols **2**, **3**, **8**, **9**, **16**, **17**, and **20** were expected to be the initial metabolities from the acylic dienes **1**, **7**, **15**, and **19**. Authentic samples of these diols were prepared by osmylation (OsO_4) of the corresponding dienes **1**, **7**, **15**, and **19** and were identified by NMR spectroscopy, GC/MS, and GC/IR analysis and by comparison with literature data. Enantiomerically enriched (7–79% *ee*) samples of these diols were also obtained by catalytic asymmetric dihydroxylation using osmium tetroxide in the presence of a chiral ligand (AD-mix-β). This chemical method of asymmetric dihydroxylation has been studied extensively by *Sharpless et al.* [18, 19] using a range of alkenes and dienes, and a model has been developed which allows tentative assignments of the absolute configurations of the product diols to be made.

In view of its importance in the environment, the biotransformation of isoprene (**1**) was studied more rigorously than that of butadiene (**19**) or the diene isomers **7** and **15** (Table 1). Using the ML2 and 8859 strains of *P. putida* as sources of BDO and NDO, respectively, and **1** as substrate, diols **2** and **3** were isolated as initial metabolites. These diols could not be detected using GC/MS analysis and the TDO enzyme present in the UV 4 strain (Table 1). The regioselectivity of dioxygenase-catalyzed dihydroxylation was consistently found to favour attack at the monosubstituted alkene bond relative to the *gem*-disubstituted alkene bond

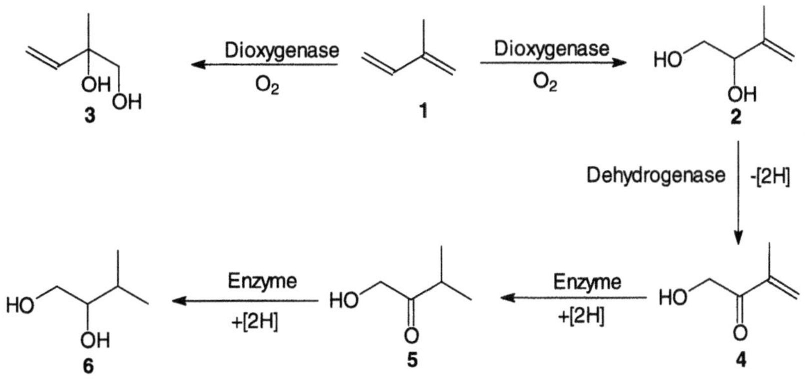

Scheme 1

Table 1. Concentrations of bioproducts **2**, **3**, **5**, and **6**, respectively, and enantiomeric excess values (*ee*) obtained from the biodegradation of isoprene (**1**) by the ML2, UV4, and 8859 strains of *P. putida*

	Propylene glycol %v/v	[**3**]/m*M*	*ee* (**3**)	[**2**]/m*M*	*ee* (**2**)	[**2**]/[**3**]	[**5**]/m*M*	[**6**]/m*M*
ML2[a]	0	1.5	40%	6.3	34%	4.2	2.0	0.2
ML2[a]	15	1.5	44%	6.4	16%	4.3	0	0
UV4[b]	0	0		0			0	0
UV4[b]	15	0.4	45%	0.9	9%	2.3	0	0
8859[c]	0	0.06	14%	0.2	12%	3.3	0	0
8859[c]	15	0.2	16%	0.6	12%	3.0	0	0

[a] BDO; [b] TDO; [c] NDO

(**2/3** = 3.3–4.2). The highest yields of diols **2** (6.3 m*M*) and **3** (1.5 m*M*) were obtained using the ML2 strain. Using the latter strain the yield of diol **2** was further diminished by oxidative metabolism under the catalytic influence of an alcohol dehydrogenase enzyme to yield ketoalcohol **4** which was further metabolized to yield the saturated ketoalcohol **5** and saturated diol **6**. Since the tertiary alcohol group in diol **3** was unaffected by the alcohol dehydrogenase, no further metabolites were found.

It had earlier been found that propylene glycol can act as a competitive inhibitor both for a diol dehydrogenase enzyme isolated from the wild-type ML2 strain (*ca.* 36% activity compared with the parent substrate, benzene *cis*-glycol) [20] and for a diol dehydrogenase present in an *E. coli* recombinant strain cloned from the ML2 strain (*ca.* 189% activity compared with the parent substrate, benzene *cis*-glycol) [21]. Hence, use of the ML2 strain and addition of propylene glycol was expected to reduce the rate of formation of ketoalcohol and to give a more reliable estimate of both regio- and stereoselectivity resulting from dioxygenase-catalyzed dihydroxylation. The effect of adding propylene glycol (15% v/v) to biotransformations of **1** with the three *P. putida* strains is evident from Table 1. Whereas the yields of diols **2** and **3** increased slightly using the wild-type ML2 and 8859 strains, ketoalcohol formation was eliminated when the ML2 strain was used with propylene glycol as additive. A more marked effect was observed with the UV4 strain where formation of **2** and **3** was increased to levels which were readily detectable (0.4 and 0.9 m*M*, respectively). This result suggests that the mutant UV 4 strain, although lacking the arene *cis*-dihydrodiol dehydrogenase enzyme, may contain a further type of dehydrogenase which can oxidize acyclic substrates like **2** and **3** and which is also effectively inhibited by propylene glycol.

The stereoselectivity of the enzyme-catalyzed dihydroxylation process was determined by use of a chiral stationary phase (CSP) GC column to resolve the enantiomers followed by identification using mass and IR spectroscopy. CSPGC/FID analysis was used to determine *ee* values of the samples of diols **2**, **3**, **8**, **9**, **16**, **17**, and **20** obtained using both chemical (OsO_4 and AD-mix-β) and biological (dioxygenase) asymmetric dihydroxylation catalysts. The absolute configurations were tentatively assigned on the basis of (*i*) the preferred configurations of diols produced using the β-form of AD-mix and the model developed by *Sharpless* for

Table 2. Enantiomeric excess, absolute configuration, and ratio of diol isomers **2**, **3**, **8**, **9**, **16**, **17**, and **20** obtained by asymmetric dihydroxylation using AD-mix-β

Diol	*ee* (%)	Abs. config.	Isomer ratio
3	21	2*R*	
2	64	2*R*	2.0 (**2/3**)
9	79	2*R*, 3*R*	
8	65	2*R*	0.36 (**8/9**)
17	7	2*R*, 1*S*	
16	44	2*R*	1.8 (**16/17**)
20	45	2*R*	

predicting the enantiofacial selectivity of the dihydroxylation process for different alkene types [18, 19] (Table 2) and (*ii*) the elution sequence observed using cyclodextrin-based CSPGC columns. The late eluting enantiomer of each diol was assigned the *R* configuration at the allylic chiral centre.

The enantiomeric excess values observed for the diols **2**, **3**, **8**, **9**, **16**, **17**, and **20** obtained by asymmetric dihydroxylation using β-AD-mix (H_2O:*t*-BuOH, 20°C) were found to be in the range of 7–79% with the lowest value being associated with dihydroxylation of a *cis*-disubstituted alkene bond as predicted from earlier studies [18, 19]. Using the BDO enzyme (from ML2), **1** was asymmetrically dihydroxylated to yield diols **2** (16% *ee* in the presence of propylene glycol) and **3** (44% *ee* in the presence of propylene glycol) with an excess of the *R* enantiomer. This enantiomeric excess increased (34% *ee*) for diol **2** when the diol dehydrogenase enzyme was present. When the TDO (from UV4) and NDO (from 8859) enzymes were used, the stereoselectivity was again generally low for diols **2** (9–12%) and **3** (14–45%) but also showed a preference for the *R* configuration.

When the TDO- and NDO-containing strains of *P. putida* (UV4 and 8859) and **1** as substrate were studied, the diols **2** and **3** were formed without evidence of further metabolism to the corresponding ketoalcohol **4** or their derivatives. Similarly, when the BDO-containing strain (ML2) was used, only a trace of ketoalcohol **4** was detected by GC/MS and GC/IR analysis; however, in this case significant quantities of the saturated ketoalcohol **5** (2.0 m*M*) and diol **6** (0.2 m*M*) were identified. The unexpected formation of the latter two metabolities was confirmed by addition of an authentic sample of the unsaturated ketoalcohol **4** to ML2 cultures. The alkene bond in compound **4** was found to be rapidly hydrogenated to yield the ketol **5** which was in turn slowly reduced to diol **6**. Further evidence for the metabolic sequence alkene dihydroxylation $\rightarrow$ allylic alcohol oxidation $\rightarrow$ alkene hydrogenation $\rightarrow$ saturated ketone reduction (Scheme 1) was obtained from a time-course study over 24 h which showed the unsaturated diol **2** to be the most abundant metabolite during the biotransformation, whereas the saturated diol **6** was only detected after 12 h (Fig. 1).

A time-course study of the biotransformation of the racemic *cis*-diol metabolites of isoprene (**2** and **3**) using the ML2 strain over a 24 h period again showed that the concentration of **3** remained constant, whereas the proportion of **2** decreased as the amounts of **5** and **6** increased (Fig. 2). CSPGC along with both FID and MS detection demonstrated that the *R* enantiomer of **2** was selectively

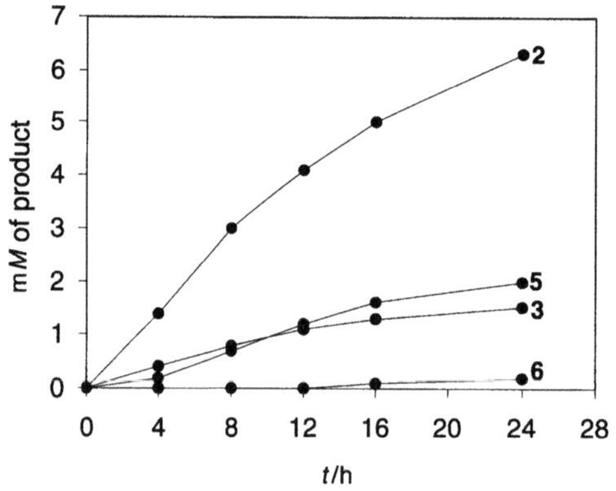

Fig 1. Time-course study of the formation of metabolites **2**, **3**, **5**, and **6** from isoprene (**1**) using *P. putida* ML2

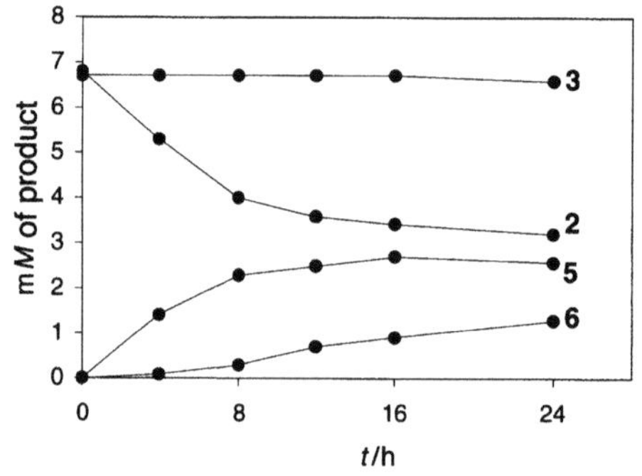

Fig 2. Time-course study of the formation of metabolites **5** and **6** from racemic diols **2** and **3** using *P. putida* ML2

oxidized to ketoalcohol **4**, and after 24 h the residual diol was mainly of the *S* configuration (95% *ee*). The enantiopurity of **2** obtained during the biotransformation of **1** is thus due to the combined effect of asymmetric dihydroxylation and kinetic resolution each favouring the *R* configuration and is an example of enantiocomplementarity. It has been established from metabolic studies of arene substrates [20] that the ML2 strain of *P. putida* contains both benzene dioxygenase and *cis*-diol dehydrogenase enzymes which could in the present context account for the formation of diols **2** and **3** and the ketoalcohol **4**. The nature of ML2 enzymes responsible for the final two steps, *i.e.* hydrogenation to yield ketol **5** and ketone reduction to yield diol **6**, has not been established. Formation of the saturated ketoalcohol **5** and saturated diol **6** indicates that enzymatic hydrogenation can occur on the *gem*-disubstituted alkene.

Based upon the higher yields of diol metabolites obtained and the unexpected results resulting from both oxidation (**2** → **4**) and reduction steps (**4** → **5** → **6**) using the ML2 strain with **1** as substrate, the related diene substrates *trans*-piperylene (**7**), *cis*-piperylene (**15**), and 1,3-butadiene (**20**) were also examined using this strain (Schemes 2–4). In accordance with expectations, the *trans* isomer **7** was found to yield the corresponding diol metabolites **8** (0.6 m*M*) and **9** (0.3 m*M*) with regioselectivity favouring the monosubstitued alkene bond, *i.e.* the rate of

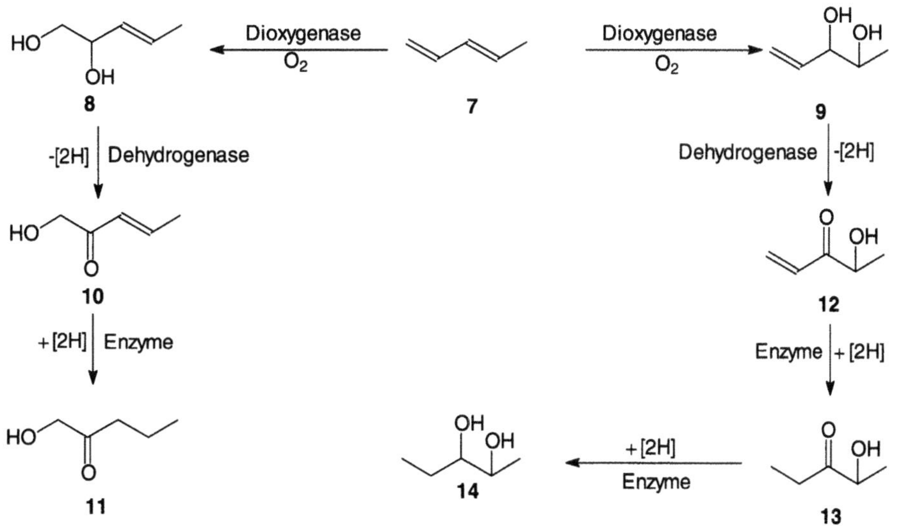

Scheme 2

formation of diol **8** was *ca.* five-fold faster compared with diol **9**. In the presence of propylene glycol, which was added to inhibit further oxidative metabolism, both diols were found to be formed in relatively low enantiomeric excess (**8**, 38% *ee*; **9**, 33% *ee*). Using the methods discussed earlier, the absolute configurations were tentatively assigned as 2*R*, 3*R*, and 2*R* for diols **9** and **8**, respectively. In the absence of propylene glycol, diol **8** was rapidly metabolized in a stereoselective manner to yield the unsaturated ketoalcohol **10** (0.5 m*M*) and the saturated ketoalcohol **11** (1.6 m*M*). The residual diol **8** was found to be of high enantiopurity (93% *ee)* and of the 2*S* configuration. In a similar manner, diol **9** was further biotransformed but at a slower rate (presumably *via* the ketoalcohol **12** which was undetected) to the saturated ketoalcohol **13** (0.1 m*M*) and the diol **14** (0.1 m*M*). The formation of the saturated ketoalcohols **11** and **13** and of the saturated diol **14** indicates that the enzyme-catalyzed hydrogenation can occur on both monosubstituted and *trans*-disubstituted alkenes. The remaining diol **9** was found to be present with an excess of the 2*S*, 3*S* enantiomer (15% *ee*). Evidence in support of the metabolic sequences **7** → **8** → **10** →**11** and **7** → **9** → **12** → **13** → **14** was obtained by addition of the

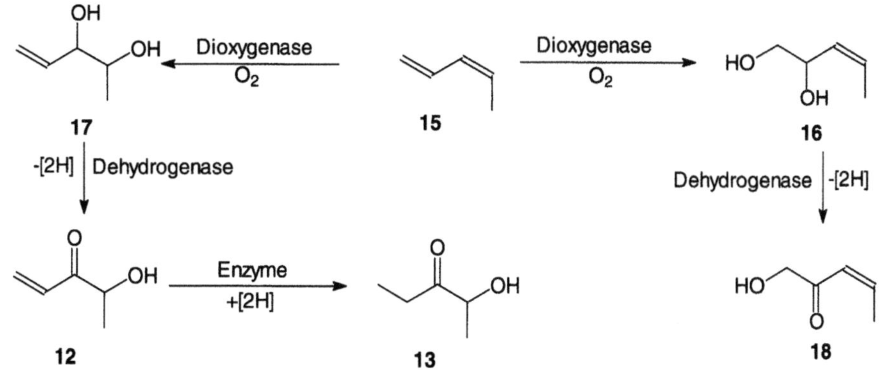

Scheme 3

racemic diols **8** and **9** individually as substrates and the detection of the corresponding metabolites **10**, **11**, **13**, and **14**. The formation of both diols **8** and **9** with an excess of the *R* allylic alcohol chiral centre from asymmetric dihydroxylation (in the presence of an inhibitor) and the preferential oxidation of this enantiomer to leave mainly the *S* enantiomer after kinetic resolution again provides an example of enzymatic enantiocomplementarity, *i.e.* the use of enzymes to produce either enantiomer.

The biotransformation of *cis*-1,3-pentadiene (**15**) using the ML 2 strain in the presence of propylene glycol yielded the expected diol metabolites **16** (1.1 m*M*; 74% *ee*) and **17** (1.2 m*M*; 70% *ee*). As before, the dioxygenase-catalyzed dihydroxylation appeared to favour diol formation at a monosubstituted alkene bond over a disubstituted alkene bond (*ca.* 3-fold rate difference) and of an enantiomer having an *R* configuration at the allylic chiral centre. Preferential removal of the allylic *R* enantiomer from diols **16** and **17** was again observed in the absence of propylene glycol inhibitor, leading to a residual sample of diol **16** (26% *ee*, 2*R*) and diol **17** (97% *ee*, 2*S*, 3*R*). Diol **16** proved to be a particularly good substrate yielding only the unsaturated ketoalcohol product **18** (4.4 m*M*), whereas diol **17** yielded the saturated ketoalcohol **13** (1.0 m*M*), probably *via* reduction of the saturated ketoalcohol **12**. It is noteworthy that while enzymatic hydrogenation was found on the monosubstituted alkene group in the ketoalcohol **12**, the *cis*-disubstituted alkene group in metabolite **18** was not hydrogenated.

On account of its volatility, butadiene (**19**) proved to be more difficult to study experimentally using the ML 2 strain of *P. putida*. However, by bubbling gaseous **19** into heptamethylnonane (0°C) and adding the resulting solution to the culture medium (4°C), dioxygenase-catalyzed dihydroxylation was found to yield diol **20** (0.2 m*M*) with a relatively low preference for the 2*R* enantiomer (25% *ee*). In the absence of the diol dehydrogenase inhibitor the 2*R* enantiomer of diol **20** was selectively oxidized to yield the saturated ketoalcohol **22** (0.3 m*M*). The unreacted diol **20** was found to be mainly of 2*S* configuration (95% *ee*). Thus, the enantiocomplementarity previously observed during the biotransformations of dienes **1**, **7**, and **15** to yield the corresponding chiral diols was again observed for diene **19**. It is noteworthy that the biotransformation of monosubstituted benzene and naphthalene substrates by *P. putida* mutant strains has been reported [22] to yield a single *cis*-dihydrodiol enantiomer and that the closely coupled *cis*-diol dehydrogenase was found to act exclusively on the same enantiomer.

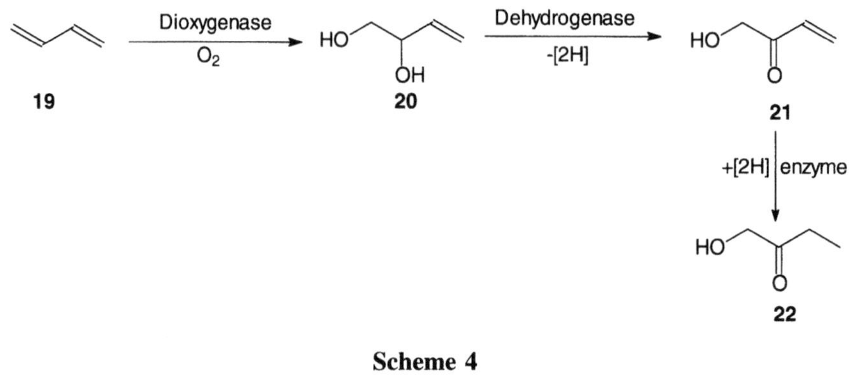

Scheme 4

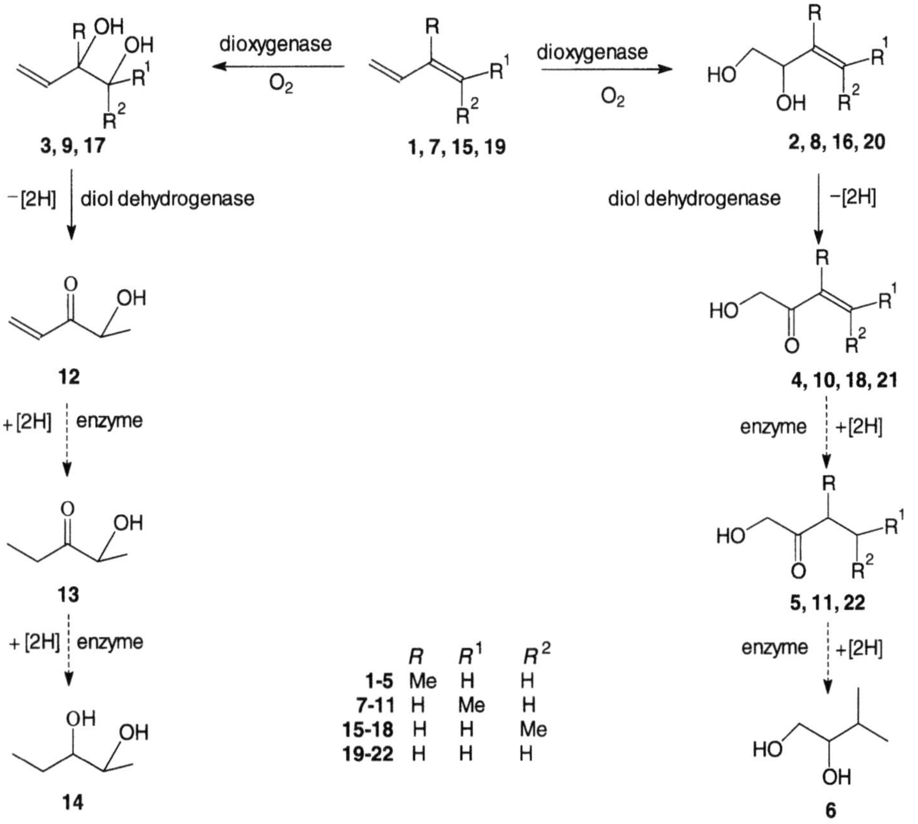

	R	*R*¹	*R*²
1-5	Me	H	H
7-11	H	Me	H
15-18	H	H	Me
19-22	H	H	H

Scheme 5

Conclusions

Bacterial biodegradation of isoprene (**1**) and related diene substrates **7**, **15**, and **19** using three different strains of the soil bacterium *P. putida* has generally been found to proceed *via* the dioxygenase-catalyzed asymmetric dihydroxylation of an alkene group to yield all of the possible diols **2**, **3**, **8**, **9**, **16**, **17**, and **20**, each with an excess of the allylic *R* enantiomer. The rate of dihydroxylation appeared to be faster on monosubstituted rather than disubstituted alkene bonds. Further diol dehydrogenase-catalyzed oxidation at the secondary allylic hydroxyl group to yield unsaturated ketoalcohols **4**, **10**, **12**, **18**, and **21** was found to occur preferentially on the allylic *R* enantiomer, thus leading to the recovery of diols having an excess of the allylic *S* configuration. The unsaturated ketoalcohol bioproducts were further metabolized through a reductive process involving enzymatic hydrogenation to yield saturated ketoalcohols **5**, **11**, **13**, and **22** and ketone reduction to yield diols **6** and **14**. A general biodegradation pathway to account for the formation of all metabolites from dienes **1**, **7**, **15**, and **19** is shown in Scheme 5.

Experimental

NMR analyzis was carried out using both Bruker Avance DPX-300 (300 MHz) and DRX-500 (500 MHz) instruments. All NMR samples were run in $CDCl_3$ using internal *TMS* as reference.

Coupling constants are expressed in Hz. GC/MS analyses were carried out using a Hewlett Packard 6890 gas chromatograph directly linked to a Hewlett Packard 5973 Mass Selective Detector (MSD). Mass spectroscopic data were obtained by operating the MSD in the electron impact scanning mode measuring ion currents between $m/z = 30$ and 400. The instrument was also operated in the chemical ionization mode using methane as reagent gas (CIMS) and measuring ion currents between $m/z = 65$ and 400 to determine the molecular ion for each compound. GC/IR analysis was completed using a Hewlett Packard 5890A gas chromatograph/5965B – infrared detector/mass selective detector (GC/IRD/MSD) system with the IRD/MSD in the series configuration. The IRD scanned the range of 750–4000 cm^{-1}. The gas chromatographs which were fitted with Supelco fused silica capillary columns (30 m×0.25 mm i.d. with a 0.25µm Beta Dex-120 phase) were equipped with autosamplers. The ovens were programmed at 65°C for 1 min and then ramped at 10°C min^{-1} to 220°C and held at this temperature for 5 min. Samples (1 mm^3) were injected in the splitless mode into the injector port which was maintained at 250°C. Helium was employed as the carrier gas at a flow rate of 1 $cm^3 \cdot min^{-1}$. Quantification of analytes was performed using gas chromatography with flame ionisation detection (GC/FID). Chromatographic separations for FID were performed using the following conditions: system A: 10 m Chrompak CP-Chirasil-DEX CB fused silica column (3.3 psi, isothermal at 100°C); system B: 10 m Chrompak CP-Chirasil-DEX CB fused silica column (3.3 psi, isothermal at 110°C); system C: 30 m Supelco-β-DEX 225 column (35 psi, isothermal at 100°C); system D: 30 m Supelco-β-DEX 225 column (19 psi; isothermal at 100°C).

The dienes **1**, **7**, **15**, and **19** (>97% purity) were purchased from Aldrich and used as substrates without further purification. Diol **20** in racemic form was available from Fluka.

Authentic samples of the known diol metabolites **2**, **3**, **8**, **9**, **16**, and **17** were obtained by osmylation. This hydroxylation procedure proved to be rather difficult due to the reactivity of the initially formed diols in the presence of OsO_4 and hence resulted in relatively low yields. In a typical general procedure a solution of diene (50 m*M*) and N-methylmorpholine-N-oxide (55 m*M*) in a mixture of acetone (50 cm^3) and water (1.5 cm^3) was added to a catalytic amount of OsO_4. After stirring overnight at ambient temperature the solution was concentrated *in vacuo*. The residual oil was dissolved in CH_2Cl_2, dried over $MgSO_4$, concentrated, and purified by flash chromatography (silica gel) starting with CH_2Cl_2 (100%) as eluent and gradually increasing the acetone concentration to 10% v/v. The product was in each case an oil containing a racemic mixture of the two isomeric diols (**2/3 8/9**, **16/17**). The yield of diols obtained were in the range of 20–30%. The structures of the individual diol metabolites were confirmed by comparison of IR, ^1H NMR, and MS data comparison with those of authentic samples. The ML 2 strain of *P. putida* was obtained from University of Warwick (Prof. *H. Dalton*); the NCIMB 8859 strain was available from the Collection of Industrial and Marine Bacteria Ltd., Aberdeen, UK. Since authentic samples of the *cis*-diols **2/3**, **8/9**, **16/17**, and **20** were available and since these were always the initial (and often the major) metabolites, a summary of the relevant spectroscopic and chromatographic data for each of these known racemic compounds is provided [4, 8].

(2R/2S)-1,2-Dihydroxy-2-methyl-3-butene (**3**; $C_5H_{10}O_2$)

^1H NMR (500 MHz, δ, $CDCl_3$): 1.3 (3H, s, Me), 3.45 (2H, d, $J = 11.0$, CH*H*OH), 3.50 (1H, d, $J = 11.0$, C*H*HOH), 3.73 (2H, bs, OH), 5.2 (1H, dd, $J = 10.4$, 1.0, =CH*H*), 5.35 (1H, dd, $J = 17.6$, 1.0, =C*H*H), 5.9 (1H, dd, $J = 10.4$, 17.6, –CH=) ppm; GC/IR: $\nu = 1644$, 3095 (H_2C=CH), 3626 (OH) cm^{-1}; GC/CIMS: $MH^+ = 103$; GC/EIMS: $m/z =$ 71 (100), 43 (82), 41 (27), 31 (18), 87 (3); CSPGC/FID: system A: 2*S*-**3** (7.8 min), 2*R*-**3** (8.1 min).

(2R/2S)-1,2-Dihydroxy-3-methyl-3-butene (**2**; $C_5H_{10}O_2$)

^1H NMR (500 MHz, δ, $CDCl_3$): 1.75 (3H, s, Me), 3.55 (1H, dd, $J = 11.0$, 7.3, *H*HCOH), 3.67 (1H, dd, $J = 11.0$, 2.8, H*H*COH), 4.18 (1H, dd, $J = 7.3$, 2.8, C*H*OH), 4.96 (1H, d, $J = 0.8$, =CH*H*), 5.06

(1H, d, $J=0.8$, =C*H*H) ppm; GC/IR: $\nu=1649$, 3085 ($C=CH_2$), 3636 (OH) cm^{-1}; GC/EIMS: *m/z* = 102 (M^+, 0.5), 71 (100), 43 (92), 41 (59), 39 (43), 31 (25), 87 (1); CSPGC/FID: system A: 2*S*-**2** (15.1 min), 2*R*-**2** (16.3 min).

(2S,3S/2R,3R)-2,3-Dihydroxy-4-pentene (**9**; $C_5H_{10}O_2$)

^1H NMR (500 MHz, δ, $CDCl_3$): 1.1 (3H, d, $J=6.3$, Me), 3.0 (2H, bs, 2×OH), 3.65 (1H, dq, $J=6.3$, 6.3, C*H*OHMe), 3.7 (1H, dd, $J=7.0$, 6.3, C*H*OH), 5.24 (1H, d, $J=10.4$, =CH*H*), 5.36 (1H, d, $J=17.0$, =C*H*H), 5.85 (1H, ddd, $J=17.0$, 10.4, 6.3, $H_2C=CH$) ppm; GC/IR: $\nu=1644$, 3090 ($H_2C=CH$), 3618 (OH) cm^{-1}; GC/EIMS: *m/z* = 45 (100), 58 (78), 57 (56), 43 (28), 31 (14); CSPGC/FID: system B: 2*S*,3*S*-**9** (4.84 min), 2*R*-3*R*-**9** (5.23 min).

Trans-(2R/2S)-1,2-Dihydroxy-3-pentene (**8**; $C_5H_{10}O_2$)

^1H NMR (500 MHz, δ, $CDCl_3$): 1.7 (3H, d, $J=6.5$, Me), 2.3 (2H, bs, 2×OH), 3.3 (1H, dd, $J=11.0$, 7.7, CH*H*), 3.7 (1H, dd, $J=5.3$, 11.0, C*H*H), 4.2 (1H, ddd, $J=8.4$, 7.0, 3.5, C*H*OH), 5.48 (1H, ddq, $J=12.6$, 6.9, 15.5, *H*C=CHMe), 5.8 (1H, dq, $J=15.5$, 6.5, =C*H*Me) ppm; GC/IR: $\nu=1672$, (C=C), 3635 (OH) cm^{-1}; GC/EIMS: *m/z* = 102 (M^+, 0.5), 71 (100), 43 (42), 41 (40), 39 (26), 53 (26), 31 (21), 87 (0.5); CSPGC/FID: system B: *S*-**8** (7.4 min), *R*-**8** (8.2 min).

(2R3S/2S3R)-2,3-Dihydroxy-4-pentene (**17**; $C_5H_{10}O_2$)

^1H NMR (500 MHz, δ, $CDCl_3$): 1.15 (3H, d, $J=6.4$, Me), 2.15 (2H, bs, 2×OH), 3.88 (1H, dq, $J=6.4$, 3.6, C*H*Me), 4.09 (1H, dd, $J=6.3$, 3.6, C*H*OH), 5.27 (1H, ddd, $J=10.5$, 1.2, 1.2, H*H*C=), 5.35 (1H, ddd, $J=17.2$, 1.2, 1.2, *H*HC=), 5.91 (1H, ddd, $J=17.2$, 10.6, 6.5, =CH) ppm; GC/IR: $\nu=1644$, 3089 ($H_2C=CH$), 3632 (OH) cm^{-1}; GC/EIMS: *m/z* = 45 (100), 58 (83), 57 (60), 43 (19); CSPGC/FID: system C: 2*S*,3*R*-**17** (9.2 min), 2*R*,3S-**17** (9.5 min).

cis-(2R/2S)-1,2-Dihydroxy-3-pentene (**16**; $C_5H_{10}O_2$)

^1H NMR (500 MHz, δ, $CDCl_3$): 1.71 (3H, d, $J=7.0$, Me), 2.15 (2H, bs, 2×OH), 3.50 (1H, dd, $J=11.2$, 7.9, C*H*HOH), 3.59 (1H, dd, $J=11.2$, 3.7, CH*H*OH), 4.59 (1H, ddd, $J=8.4$, 8.4, 3.6, C*H*OH), 5.39 (1H, m, =C*H*Me), 5.65 (1H, m, HC=) ppm; GC/IR: $\nu=1658$ (C=C), 3636 (OH) cm^{-1}; GC/EIMS: *m/z* = 71 (100), 43 (41), 41 (41), 39 (27), 53 (27), 31 (22), 87 (1); CSPGC/FID: system C: 2*S*-**16** (14.6 min) 2*R*-**16** (15.6 min).

(2R/2S)-1,2-Dihydroxy-3-butene (**20**; $C_4H_8O_2$)

^1H NMR (500 MHz, δ, $CDCl_3$): 3.50 (1H, dd, $J=11.0$, 7.5, CH*H*OH), 2.95 (2H, bs, 2×OH), 3.67 (1H, dd, $J=11.4$, 3.4, C*H*HOH), 4.26 (1H, m, C*H*OH), 5.23 (1H, ddd, $J=10.6$, 1.4, 1.4, H*H*C=), 5.35 (1H, ddd, $J=17.3$, 1.5, 1.5, *H*HC=), 5.85 (1H, ddd, $J=17.0$, 10.4, 5.6), ppm GC/IR: $\nu=1648$ (C=C), 2887, 2938, 3091 (CH), 3641 (OH) cm^{-1}; GC/EIMS: *m/z* = 57 (100), 58 (65), 31 (62), 39 (44), 70 (37), 42(35); CSPGC/FID: system D: 2*S*-**20** (10.6 min) 2*R*-**20** (11.0 min).

Metabolites **5**, **6**, **10**, **11**, **13**, **14**, **18**, and **22** were identified from IR, EIMS and CIMS data obtained directly after GC separation.

1-Hydroxy-3-methyl-2-butanone (**5**; $C_5H_{10}O_2$)

GC/IR: $\nu=3530$ (OH), 1727 (C=O) cm^{-1}; GC/EIMS: *m/z* = 102 (M^+, 1), 31 (31), 43 (100), 71 (35), 41 (48); GC/CIMS: *m/z* = 103 (MH^+).

3-Methyl-1,2-butanediol (**6**; $C_5H_{12}O_2$)

GC/IR: $\nu = 3644$ (OH), 2968, 2892 (CH), 1058 cm^{-1}; GC/EIMS: *m/z* = 73 (100), 55 (70), 43 (60), 61 (40), 31 (39), 41 (25); GC/CIMS: *m/z* = 87 ($MH-H_2O^+$).

(E)-1-Hydroxy-3-penten-2-one (**10**; $C_5H_8O_2$)

GC/IR: $\nu = 3525$ (OH), 1702 (α,β-unsatd. C=O), 1643, 966 (C=C) cm^{-1}; GC/EIMS: *m/z* = 69 (100), 41 (98), 39 (54); 31 (19); GC/CIMS: *m/z* = 101 (MH^+).

1-Hydroxy-2-pentanone (**11**; $C_5H_{10}O_2$)

GC/IR: $\nu = 3530$ (OH), 1728 C=O) cm^{-1}; GC/EIMS: *m/z* = 102 (M^+, 6), 43 (100), 71 (51), 31 (34); GC/CIMS: *m/z* = 103 (MH^+).

2-Hydroxy-3-pentanone (**13**; $C_5H_{10}O_2$)

GC/IR: $\nu = 3535$ (OH), 1728 C=O) cm^{-1}; GC/EIMS: *m/z* = 102 (M^+, 2), 45 (100), 57 (36), 31 (9); GC/CIMS: *m/z* = 103 (MH^+).

2,3-Pentanediol (**14**; $C_5H_{12}O_2$)

GC/EIMS: *m/z* = 59 (100), 31 (68), 45 (60), 41 (31), 75 (14); GC/CIMS: *m/z* = 87 ($MH-H_2O^+$).

(Z)-1-Hydroxy-3-penten-2-one (**18**; $C_5H_8O_2$)

GC/IR: $\nu = 3525$ (OH), 1701 (α,β-unsatd. C=O), 1631 (C=C) cm^{-1}; GC/EIMS: *m/z* = 100 (M^+, 2), 69 (100), 41 (79), 39 (52), 31 (21); GC/CIMS: *m/z* = 101 (MH^+).

1-Hydroxy-2-butanone (**22**; $C_4H_8O_2$)

GC/IR: $\nu = 3536$ (OH), 1733 (C=O) cm^{-1}; GC/EIMS: *m/z* = 88 (M^+,13), 57 (100), 31 (36); GC/CIMS: *m/z* = 89 (MH^+).

Substrates were metabolized on a small scale using shake flask cultures of the constitutive mutant strain (UV4) of *P. putida* and the wild type *P. putida* NCIMB 8859 according to reported methods [15, 16, 20]. The wild type *P. putida* ML2 was grown on benzene as the sole carbon source. To obtain biomass, a 10 litre Biolab fermenter containing rich media (Luria Broth) was inoculated with a 1% inoculum of benzene grown cells. The cells were harvested by centrifugation (10000 g for 15 min) during the late exponential phase of growth and resuspended in potassium phosphate buffer (100 m*M*, *pH* = 7.5) to an optical density of 10 at 600 nm. These cell suspensions were used for the bioconversions. A typical procedure involved incubation of diene substrates **7**, **15**, and **19** (60 mm^3 in 3 cm^3 of culture medium containing the bacterial cells, *OD* = 10 at 600 nm) in shake flasks (10 cm^3) at a temperature of +4°C and *pH* = 7.5 for 24 h. As dienes **7**, **15**, and **19** had a very low solubility in water, a two-phase system was formed with the aqueous phase being saturated with the diene. In the case of 1,3-butadiene (**19**) the substrate was passed as a gas into heptamethylnonane to yield a saturated solution at 0°C. Subsequently, a portion (200 mm^3) of this solution was added to the bacterial cell solution. Bioconversions with the chemically synthesized mixtures of racemic diols **2**, **3**, **8**, **9**, **16**, **17**, and **20** as substrates for *P. putida* ML2 were carried out at a diol concentration of 2.0 $g.dm^{-3}$. The metabolites were obtained by rigorous extraction of the culture medium with ethyl acetate after saturation with salt. The extract was concentrated and analyzed by chiral stationary phase FID, MS, and IR detection. The structures of the *cis*-diol metabolites **2**, **3**, **8**, **9**, **16**, **17**, and **20** were established by comparison with authentic samples. The remaining ketoalcohols (**5**, **10**, **11**, **13**,

18, and **21**) and diols (**6** and **14**) were identified on the basis of comparison with literature data and from MS and IR spectra.

Note Added in Proof

In a recent report [23] the metabolism of isoprene in a *Rhodococcus* strain was found to occur *via* a monooxygenase-catalyzed epoxidation. No evidence of diene epoxidation was obtained using the *P. putida* strains in the present study.

Acknowledgements

We wish to thank Professor *H. Dalton* for the *P. putida* ML2 strain, Dr *N. I. Bowers* for the preliminary studies on *cis*-piperylene, and the Technology Development Programme of the European Regional Development Fund and the IRTU for financial support (*MC, DC*).

References

[1] Atkinson R, Arey J (1998) Acc Chem Res **31**: 574
[2] Cleveland CC, Yavitt JB (1998) Appl Environ Microbiol **64**: 172
[3] Konig G, Brunda M, Puxbaum H, Hewitt CN, Duckham SC, Rudolf J (1995) J Atmos Environ **29**: 861
[4] Becker KH, Barnes I, Ruppert L (1994) Physico-Chemical Behaviour of Atmospheric Pollutants **1**: 54
[5] Peeters A, Ectors A, Boullart W (1994) Physico-Chemical Behaviour of Atmospheric Pollutants **1**: 61
[6] Madronich S, Calvert JG (1990) J Geophys Res **95**: 5697
[7] Bleasdale C, Small RD, Watson WP, Wilson J, Golding BT (1996) Toxicology **113**: 290
[8] Wistuba D, Weigand K, Peter H (1994) Chem Res Toxicol **7**: 336
[9] Longo V, Citti L, Gervasi PG (1985) Toxicology Letters **29**: 33
[10] Gervasi PG (1985) Mutat Res **156**: 77
[11] Allen CCR, Boyd DR, Dalton H, Sharma ND, Brannigan I, Kerley NA, Sheldrake GN, Taylor ST (1995) J Chem Soc Chem Commun 117
[12] Allen CCR, Boyd DR, Hempenstall F, Larkin MJ, Sharma ND (1999) Appl and Environ Microbiol **65**: 1335
[13] Boyd DR, Sharma ND, Agarwal RA, Resnick SM, Schocken MJ, Gibson DT, Sayer JM, Yagi H, Jerina DM (1997) J Chem Soc Perkin Trans 1, 1715
[14] Boyd DR, Sharma ND, Evans TA, Groocock MG, Malone JF, Stevenson PJ, Dalton H (1997) J Chem Soc Perkin Trans 1, 1879
[15] Bowers NI, Boyd DR, Sharma ND, Kennedy MA, Sheldrake GN, Dalton H (1998) Tetrahedron Asymm **9**: 1831
[16] Boyd DR, Sharma ND, Haughey SA, Kennedy MA, McMurray BT, Sheldrake GN, Allen CCR, Dalton H, Sproule K (1998) J Chem Soc Perkin Trans 1, 1929
[17] Lange CC, Wackett LP (1997) J Bacteriology **179**: 3858
[18] Kolb HC, VanNiewenhze MS, Sharpless KB (1994) Chem Rev **94**: 2483
[19] Wang ZM, Kakiuchi K, Sharpless KB (1995) J Org Chem **59**: 6897
[20] Allen CCR, Boyd DR, Dalton H, Sharma ND, Walker CE (in preparation)
[21] Fong KPY, Goh CBH, Tan, H-M (1996) J Bacteriol **178**: 5592
[22] Allen, CCR, Boyd DR, Dalton H, Sharma ND, Brannigan I, Kerley NA, Sheldrake GN, Taylor SC (1994) J Chem Soc Chem 117
[23] van Hylckama Vlieg JE, Leemhuis H, Spelberg JH, Janssen DB (2000) J Bacteriol **182**: 1956

Received December 20, 1999. Accepted (revised) February 7, 2000

Enantioselective Sulfoxidations Catalyzed by Horseradish Peroxidase, Manganese Peroxidase, and Myeloperoxidase

Antonin Tuynman[1], **Hans E. Schoemaker**[2], and **Ron Wever**[1,*]

[1] E.C. Slater Institute, BioCentrum, University of Amsterdam, NL-1018 TV Amsterdam, The Netherlands
[2] DSM Research, Bio-Organic Chemistry, NL-6160 MD Geleen, The Netherlands

Summary. Horseradish peroxidase (HRP), myeloperoxidase (MPO), and manganese peroxidase (MnP) have been shown to catalyze the asymmetric sulfoxidation of thioanisole. When H_2O_2 was added stepwise to MPO, a maximal yield of 78% was obtained at *pH* 5 (*ee* 23%), whereas an optimum in the enantiomeric excess (32%, (*R*)-sulfoxide) was found at *pH* 6 (60% yield). For MnP a yield of 18% and a high enantiomeric excess of 91% of the (*S*)-sulfoxide were obtained at *pH* 5 and a yield of 36% and an *ee* of 87% at *pH* 7.0. Optimization of the conversion catalyzed by horseradish peroxidase at *pH* 7.0 by controlled continuous addition of hydrogen peroxide during turnover and monitoring the presence of native enzyme as well as of intermediates I, II, and III led to the formation of the sulfoxide in high yield (100%) and moderate enantioselectivity (60%, (*S*)-sulfoxide).

Keywords. Enantioselective sulfoxidation; Horseradish peroxidase; Myeloperoxidase; Manganese peroxidase.

Introduction

General

The use of enzymes in oxidative conversions is a potentially attractive method for the synthesis of optically active compounds. In this way oxidative processes employing stoichiometric amounts of heavy metal salts can be avoided, whereas the high regio- and/or enantioselectivity of enzymatic processes is a promising feature which may be exploited in various ways. Other oxidative bioconversions involve the use of whole cells, making the whole process rather cumbersome and being therefore often restricted to conversions leading to compounds with a high added value like pharmaceuticals [1]. Peroxidases, however, might be potentially attractive biocatalysts for fine chemicals production since cosubstrate recycling procedures are not necessary and hydrogen peroxide is used as a clean and cheap oxidant.

Various peroxidases have been shown to catalyze the formation of a number of alkyl arylsulfoxides or dialkylsulfoxides in an enantioselective manner [2–5]. The

* Corresponding author

compounds are useful intermediates in organic synthesis, both as chiral auxiliaries in catalytic processes and as stoichiometric intermediates. Also epoxidations [6, 7], benzylic [8] and propargylic [9] hydroxylation, and indole oxidations [10] are catalyzed by peroxidases. The potential application in organic synthesis is still hampered by the moderate stability of the peroxidases under turnover conditions, the limited solubility of organic reactants in water, and the oxidative inactivation of the heme. The catalytic performance of peroxidases may be improved by continuous controlled addition of hydrogen peroxide [11, 12], by variation of *pH* [4], and by the addition of some organic solvents [11]. We recently have shown that under carefully choosen conditions also lactoperoxidase (a commercially available enzyme) is capable of enantioselective sulfoxidation of thioanisole, affording the (*R*) isomer in a yield of 85% with an *ee* of 80% [5]. We have also shown that the opposite antipode, the (*S*)-sulfoxide, could be obtained in a yield of 84% with an *ee* of 73% using the commercially available fungal peroxidase from *Coprinus cinereus*. In addition, we have described a spectroscopic method to monitore the active enzyme intermediates during turnover. In this way we were able to optimize the conditions for maximum yield and optimum *e.e.* [5].

The aim of the present research was twofold. First we investigated if we also could use our spectroscopic method to further optimize the sulfoxidation of thioanisole to form the (*S*)-sulfoxide in high yield and high *ee* using horseradish peroxidase. Second, we investigated two other peroxidases (myeloperoxidase and manganese peroxidase) to see if these enzymes could also be used in enantioselective sulfoxidation reactions. Myeloperoxidase was expected to give the (*R*)-sulfoxide due to its structural ressemblance with lactoperoxidase [13]. Moreover, in analogy to the heme CPO it is able to oxidize chloride to hypochlorous acid [14]. The manganese peroxidase is also a fungal enzyme like CiP, and it was expected that with this enzyme – the mnp-gene has recently been cloned and expressed [15] – we would obtain the (*S*)-sulfoxide.

Mechanistic considerations

The general mechanism of peroxidases starts by the addition of H_2O_2 to the enzyme in its resting state, the native Fe^{3+} state [16]. Upon release of a molecule of water, an oxoferryl species ($Fe(IV){=}O\ P^{+\cdot}$) called compound I is created containing two oxidative equivalents, one of which is located in the porphyrin or on a protein residue as a radical-cation. The substrate is oxidized to a substrate radical in this step. Addition of an equivalent of a traditional peroxidase substrate, a one-electron donor, will reduce compound I to a $Fe(IV){=}O$ species that is one oxidative equivalent above the resting state, the so-called compound II, which does not contain an electron hole an either porphyrin or protein anymore. A second equivalent of substrate will reduce this species back to the native enzyme upon the release of a water molecule and a second substrate radical. The enzyme can also return directly from compound I back to the native enzyme upon reaction with a second molecule of H_2O_2, releasing a molecule of oxygen and a molecule of water. This is called the catalase activity of the enzyme. Another way for the direct reconversion of compound I to the native enzyme proceeds *via* oxygen-transfer to a suitable substrate such as a sulfide or an alkene (Scheme 1). This is called the oxo-

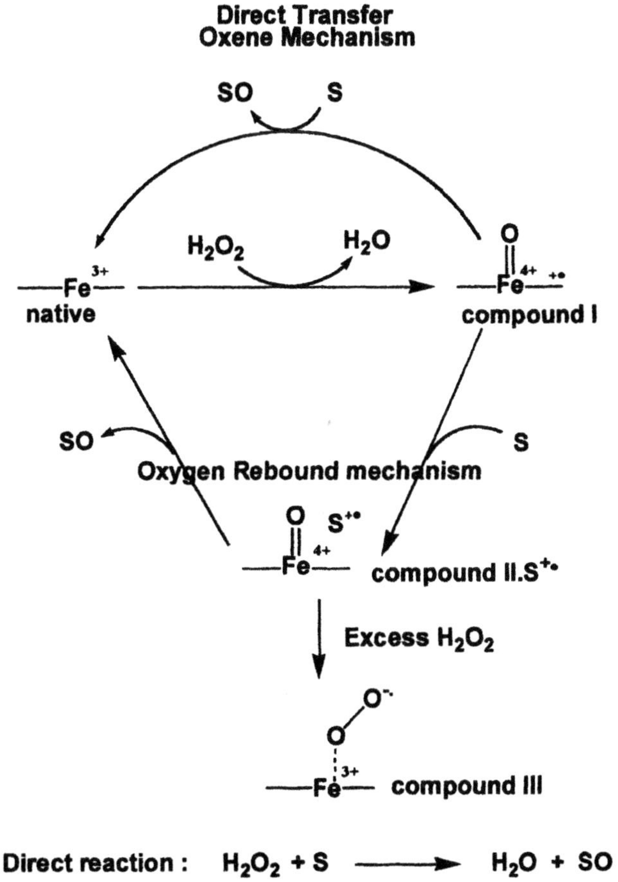

Scheme 1. Oxene and oxygen-rebound mechanisms for the oxygen transfer to sulfides; the formation of compound III under a large excess of H_2O_2 is indicated as well as the direct racemic reaction between H_2O_2 and the sulfide

ferryl or oxene mechanism [17]. Alternatively, oxygen-transfer may take place *via* a two-step mechanism: in this so-called oxygen-rebound mechanism [17–22] a molecule of substrate is first oxidized by compound I to a substrate radical-cation that forms a complex with compound II (compound II $\cdot$ $S^{+\cdot}$ Scheme 1). Subsequently the oxygen of compound II is transferred to this substrate radical-cation, a molecule of oxygenated product is released, and the enzyme returns to its native state. In the case of an oxygen-rebound mechanism the rate of transition from compound I to compound II should depend linearly on the sulfide concentration, whereas in an oxene mechanism the formation of compound II should not be observed. As we have shown, lactoperoxidase and myeloperoxidase operate *via* an oxygen-rebound mechanism [5].

At high H_2O_2 concentrations a third intermediate of the enzyme, compound III, can be formed (Scheme 1). In most reactions this intermediate is believed to be a catalytic inactive, mostly a dead-end species and leading to inactivated enzyme for most peroxidases [23]. However, compound III is not necessarily irreversible formed and may return to the native state [23]. Under operational conditions, involving discontinuous addition of hydrogen peroxide [3], compound III may well be transiently formed. It is of importance to avoid the presence of compound III, since it is indicative of a high H_2O_2 concentration [5]. This will lead to an enhanced racemic direct reaction between the sulfide and H_2O_2 and therefore lower the *ee* of the sulfoxides formed [5].

Results and Discussion

Several studies have been published on the catalysis of asymmetric sulfoxidation by horseradish peroxidase [3, 4, 24–28]. Yet there is no agreement on the best method to perform these incubations, and therefore the yield and the enantiomeric excess of the sulfoxide formed differ from report to report. This may also be due to the experimental conditions used in these studies. Differences include amount of enzyme, concentration of the sulfide, open reaction vessels *vs.* closed ones, size headspace to prevent evaporation of the substrate, *pH* values [3, 4, 24–28], and the way H_2O_2 is added (in one step [24] or stepwise [3, 4, 25–28]).

For horseradish peroxidase, we have attempted to optimize the sulfoxidation with respect to the yield whilst maintaining a high enantioselectivity. Therefore, the method previously described in detail for CiP and LPO [5] of continuously adding H_2O_2 and simultaneously monitoring the enzyme intermediates, in particular compound III, was applied. This method guarantees that no excessive accumulation of H_2O_2 will occur. The rate of addition of H_2O_2 is to be lowered when a small amount of compound III is formed, thus minimizing the direct racemic reaction between H_2O_2 and the sulfide. The reactions were carried out at $pH = 7.0$, since at that *pH* value the highest *ee* for this conversion had been reported. Initially it was found that by incubating 19.2 μ*M* HRP with 1.7 m*M* thioanisole at *pH* 7.0 at a H_2O_2 influx of 1 μmol/h, after 5 hours a small amount of compound III could be detected in the mixture of compound I and II. Therefore, the rate of H_2O_2 addition was lowered to 0.4 μmol/h, and consequently only compound I and II were present. The presence of a mixture of compounds I and II is indicative of an oxygen-rebound mechanism which could be confirmed by stopped-flow studies [21, 22]. Nonetheless, during the incubations the absorbance of the *Soret* band decreased. About 20 μ*M* enzyme were necessary to fully convert the substrate before the enzyme was inactivated.

After 48 hours all of the thioanisole was converted into the sulfoxide with an enantioselectivity of 60% of the (*S*)-enantiomer. Variation of the substrate or the enzyme concentration did not lead to an altered *ee* provided that the reaction was quenched as soon as compound III started to be formed.

It is intriguing that the enantioselectivity does not change when the substrate concentration is changed. *Kobayashi et al.* have shown that 10% of the oxygen incorporated in the sulfide derives from water [17]. It is believed that the incorporation of an oxygen atom from water occurs when a sulfide radical-cation recombines with a second molecule of radical-cation to give a molecule of sulfide and a sulfide dication; the latter will react with water. The fact that the *ee* is independent on the sulfide concentration implies that these processes take place in the active site. Indeed, the stopped-flow experiments of *Dunford et al.* are indicative of the presence of two molecules of substrate in or near the active site [22].

From the decrease in the absorption at 280 nm we determined an initial turnover frequency of $0.3\,min^{-1}$. This is significantly lower than the turnover frequency of $3\,min^{-1}$ determined by *Savenkova et al.* [29, 30]. The turnover number obtained with our optimization technique is 52, which is significantly higher than those previously reported [3, 4]. It must be mentioned that we did not use sulfide concentrations exceeding the solubility limits and that the reaction vessels were completely filled and sealed to prevent evaporation of the sulfide.

Using less favourable conditions resulted in lower yields and/or lower *ee*s. When the reactions were carried out by stepwise addition of H_2O_2 and were quenched after one hour [25–28], the yields did not exceed 11% (results not shown). After quenching, the enzyme is still active, and higher yields could have been obtained if the procedure would have been carried out for a longer time. In Refs. [25–28] higher enantioselectivities were claimed, but the amont of sulfoxides formed in a parallel reaction without enzyme was subtracted. This is not correct since the H_2O_2 concentration in the enzyme catalyzed reaction is much lower than in the absence of enzyme due to its catalase activity. In a later publications of the same group the subtraction of the racemic parallel reaction has been omitted, leading to enantioselectivities of 58% [29, 30].

When the reactions were carried out as described by *Morishima et al.* [24], *i.e.* by a single addition of 1.0 m*M* H_2O_2 at the start of the reaction, a yield of only 4% of the (*S*)-sulphoxide was obtained in one hour with an *ee* of only 52%. The low *ee* is probably due to a significant contribution of the direct reaction. Figure 1 shows that using this method the conversion of thioanisole only takes place during the first

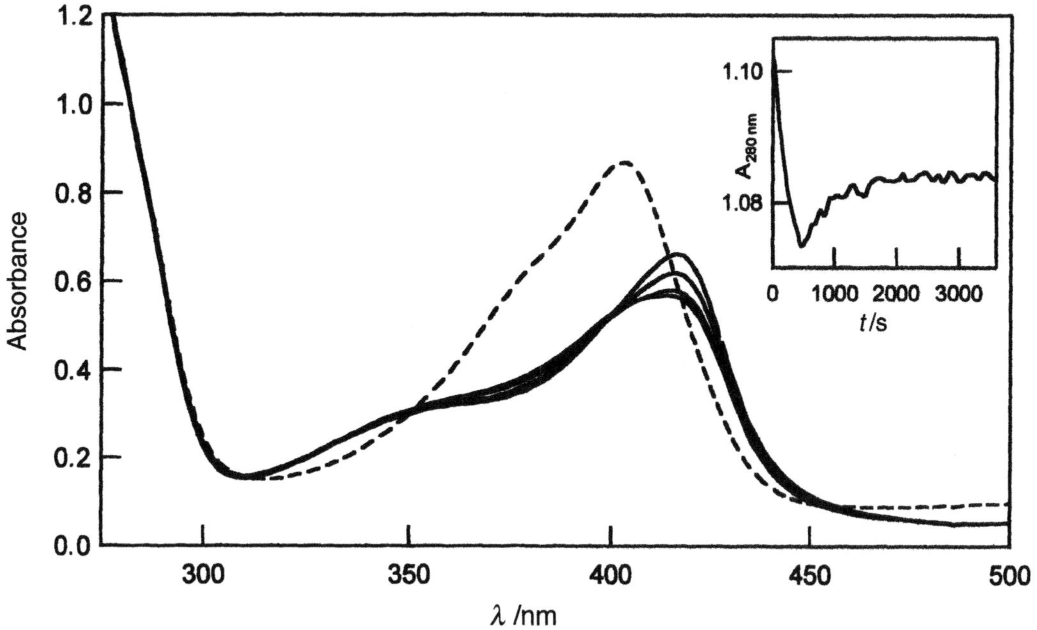

Fig. 1. Spectra of the enzyme intermediate of HRP in the sulfoxidation of thioanisole after 10, 130, 250, and 490 s (——); hereafter, the enzyme starts to return to its native state, the spectrum of which is represented after one hour (- - - -); in the insert the conversion of thioanisole is shown by the decrease in absorption at 280 nm; see experimental procedures for details

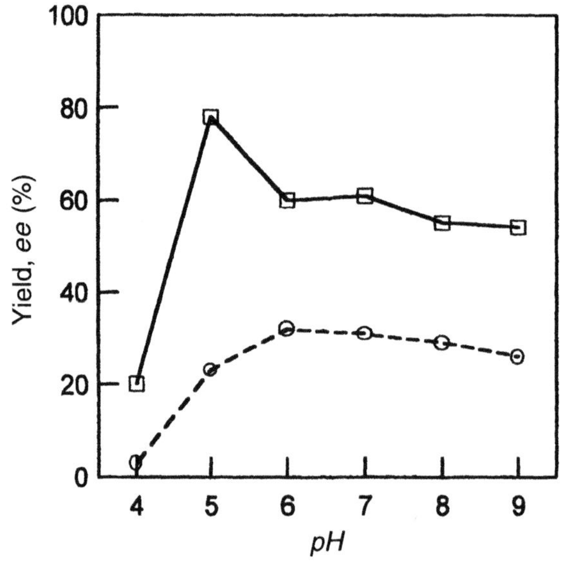

Fig. 2. *pH* dependence of the formation of (*R*)-methyl phenyl sulfoxide and its *ee* by MPO (O: *ee*, □: yield); see experimental procedures for details

500 seconds as judged from the decrease in absorption at 280 nm (insert). The following slower increase in absorption is probably due to conversion of compound I or II into native enzyme: the enzyme is mostly present as compound II and starts to return to its native state after the first 500 seconds.

In conclusion, *ee* and yield of the sulfoxidation of thioanisole by HRP depend strongly on the conditions and the design of the experiment. It was possible to convert the thioanisole completely into the sulfoxide using a slow continuous influx of H_2O_2 (on average 0.46 µmol/h) for 48 h. This influx is considerably slower than that for the same reaction catalyzed by LPO and CiP [5]. Continuous addition of H_2O_2 only is not sufficient to obtain a high *ee* it is also imperative to follow the enzyme intermediates spectroscopically to prevent the accumulation of H_2O_2 and the formation of catalytically inactive species as *e.g.* compound III.

The sulfoxidation of thioanisole by MPO was carried out *via* the traditional method of adding H_2O_2 stepwise. The *pH* dependence of the enantioselectivity and the yield are presented in Fig. 2. Although the yield is maximal at *pH* 5 (78%), the best enantioselectivity is obtained at *pH* 6.0 (32%). At higher *pH* values the enantiomeric excess remains fairly constant. The yield decreases showly at increasing *pH* values. These enantioselectivities are much higher than previously reported (4–8%) for the sulfoxidation of *p*-methylthioanisole by MPO [31]. In these experiments, much lower enzyme and substrate concentrations have been used. In agreement with Ref. [31] the rate for this reaction is maximal at *pH* 5.0. For MPO we have not optimized the yield *via* our method of controlled addition of H_2O_2 and monitoring at the same time the enzyme intermediates present. The high yields obtained employing the traditional stepwise addition of H_2O_2 are in contrast to our findings with other heme peroxidases.

Here we also report for the first time the enantioselective sulfoxidation by MnP. Due to a limited amount of purified enzyme available we were only able to carry out the experiment at *pH* 5.0 and 7.0. Best results were obtained at *pH* 5.0. Although the yield was very low (18%), the enantioselectivity was very high (91% *ee* of (*S*)-sulfoxide). This is the highest value reported for a heme peroxidase producing the (*S*)-sulfoxide and significantly higher than reported for structurally related enzymes [32] (CiP: 73% *ee* HRP: 60% *ee*). At *pH* 7.0 the yield was considerably higher (36%), but the *ee* dropped to 87%. CPO also produces the sulfoxide with a high *ee* of even 99% [11, 12], but in this case the (*R*)-enantiomer is formed.

Unfortunately, the very limited amount of MnP available did not permit us to optimize this reaction with respect to *pH* value and controlled continuous addition of H_2O_2. Small genetic modifications for MnP (such as the F41L mutation for HRP [25]) might enhance the *ee* and make this enzyme one of its mutants a promising catalyst for the production of (*S*)-sulfoxides.

Experimental

MPO was purified from human leukocytes as described ($A_{428nm/280nm} = 0.8$) [33]. Enzyme concentrations were determined using a molar extinction coefficient of 102 m$M^{-1} \cdot cm^{-1}$ at 403 nm for HRP [34], 89 m$M^{-1} \cdot cm^{-1}$ at 428 nm for MPO [33], and 127 m$M^{-1} \cdot cm^{-1}$ at 406 nm for MnP [35]. Absorption spectra in the UV/Vis range were recorded on a Hewlett Packard 8452 A spectrophotometer. Figure 3 shows the spectra of native MPO, MPO-I, MPO-II, and MPO-III. The spectra of native HRP, HRP-1, HRP-II, and HRP-III are shown in Fig. 4.

To determine the *pH* optimum for the sulfoxidation of thioanisole by MPO, reactions were carried out in 100 m*M* buffer at 25°C. Buffers used were sodium acetate (*pH* = 4.0 and 5.0), potassium phosphate (*pH* = 6.0–8.0), and sodium carbonate (*pH* = 9.0).

The reaction mixture contained 10 μ*M* MPO and 1 m*M* thioanisole. During one hour, 20×10 mm^3 of 10 m*M* H_2O_2 were added. The same procedure was applied to the reaction with MnP as a catalyst, however only at *pH* = 5.0 and 7.0 due to the very limited amount of enzyme available. Kinetic experiments following the sulfoxidation of thioanisole by HRP were conducted using a HP 8452 A spectrophotometer at 280 nm. The difference of the molar extinction coefficients of methyl phenyl sulfide and the methyl phenyl sulfoxide was determined to be 0.84 m$M^{-1} \cdot cm^{-1}$ at 280 nm. The amounts of compound I, II, or III and of native enzyme were followed simultaneously. Hydrogen

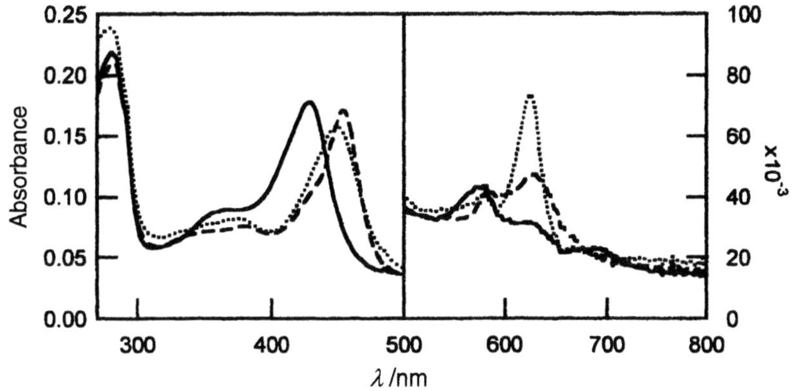

Fig. 3. Optical absorption spectra of MPO, native enzyme, compound II, and compound III; ——: 2 μ*M* MPO in 100 m*M* potassium phosphate (*pH* 7.0), - - - -: MPO-II generated by addition of 40 μ*M* hydrogen peroxide to 2 μ*M* MPO, ······: MPO-III generated by addition of 4 m*M* hydrogen peroxide to 2 μ*M* MPO

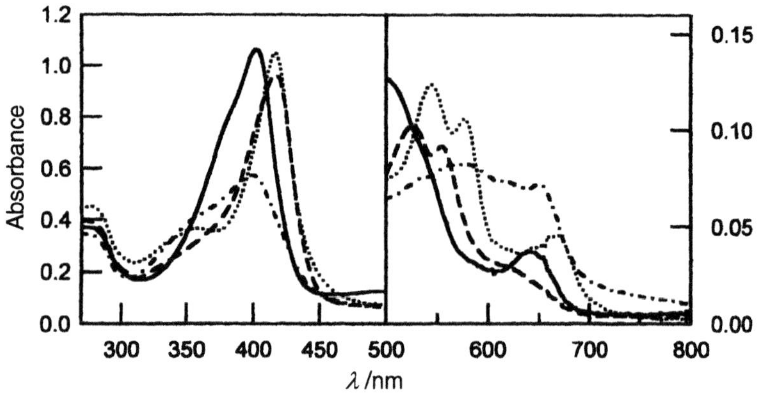

Fig. 4. Optical absorption spectra of native HRP, HRP-I, HRP-II, and HRP-III; ——: 10 μ*M* HRP in 100 m*M* potassium phosphate (*pH* 6.5), ·-·-·-·-·-·: HRP-I generated by addition of 10 μ*M* hydrogen peroxide, - - - - - -: HRP-II generated by addition of 10 μ*M* ferrocyanide to HRP-I, · · · · ·: HRP-III generated by addition of a large excess hydrogen peroxide (2.5 m*M*)

peroxide was added continuously *via* a syringe pump (Cole Parmer 74900-10) with a 250 mm^3 Hamilton syringe with a teflon luer lock connected to the reaction cuvet *via* a PEEK tubing 1/16″OD/0.20″ID that went through a capillary in the teflon cap that sealed the cuvet.

For HRP, the average rate of hydrogen peroxide addition was 0.46 μmol/h. Typically, reactions were carried out in a 1.67 cm^3 quartz cuvet sealed with a teflon cap with two capillaries: one to add the hydrogen peroxide solution and one to dispose of the overflow. The cuvets were completely filled with the reaction mixture in order to prevent partitioning of the methyl phenyl sulfide into a gas-phase headspace. The contents of the cuvet were stirred continuously. General conditions were 10 μ*M* enzyme, 100 m*M* potassium phosphate buffer, and 0.15–1.7 m*M* methyl phenyl sulfide.

The reactions were quenched with sodium sulfite [3] to consume the excess of H_2O_2 after 1 h for MPO and MnP and for HRP when the sulfoxidation process had ended as judged from a constant absorbance at 280 nm.

1 μmol of acetophenone was added as internal standard, and the reaction mixture was extracted twice with 3.4 cm^3 CH_2Cl_2. The organic layer was concentrated under a stream of nitrogen to a volume of about 20 mm^3 and diluted with 1 cm^3 of 80% hexane/20% isopropyl alcohol. A 20 mm^3 sample was loaded onto a chiracel OD HPLC column (Daicel Chemical Industries, 0.46 cm×25 cm) equipped with a Pharmacia LKB-HPLC pump 2248 and a LKB Bromma 2140 rapid spectral detector connected to a PC. The Borwin™ program was used to evaluate peak areas. The (*R*)- and (*S*)-methyl phenyl sulfoxides were eluted isocratically with the same solvent mixture at a flow rate of 0.5 ml/min and detected at 254 nm. Retention times of methyl phenyl sulfide, acetophenone, and the (*R*)- and (*S*)-methyl phenyl sulfoxides were 8.4, 9.4, 15.0, and 17.8 min, respectively [5].

H_2O_2 solutions were freshly prepared by dilution of a 30% stock solution (Merck). The concentration was determined spectrophotometrically using an absorption coefficient of 43.6 $M^{-1} \cdot cm^{-1}$ at 240 nm [36]. All other chemicals were of the highest purity. Methyl phenyl sulfide was purchased from Fluka, methyl phenyl sulfoxide and acetophenone from Aldrich.

Acknowledgements

We wish to thank Prof. *M. H. Gold* from the Oregon Graduate Institute of Science and Technology for a gift of manganese peroxidase. This work was supported by the BioCentrum and the Innovation Oriented Research Programs Catalysis (IOP catalysis). We also received financial support from the Council for Chemical Sciences of the Netherlands Organisation for Scientific Research (NWO) and the Netherlands Technology Foundation (STW).

References

[1] Holland HL (1992) In: Holland HL (ed) Organic synthesis with oxidative enzymes. VCH, New York
[2] Colonna S, Gaggero N, Manfredi A, Casella L, Gulotti M, Carrea G, Pasta P (1990) Biochemistry **29**: 10465
[3] Colonna S, Gaggero N, Carrea G, Pasta P (1992) J Chem Soc Chem Commun 357
[4] Colonna S, Gaggero N, Richelmi C, Carrea G, Pasta P (1995) Gaz Chim Ital **125**: 479
[5] Tuynman A, Vink MKS, Dekker HL, Schoemaker HE, Wever R (1998) Eur J Biochem **258**: 906
[6] Allain AJ, Hager LP, Deng L, Jacobsen EN (1996) J Am Chem Soc **117**: 4415
[7] Tuynman A, Lutje Spelberg J, Kooter IM, Schoemaker HE, Wever R (2000) J Biol Chem **275**: 3025
[8] Miller VP, Tschirret-Guth, RA, Ortiz de Montellano PR (1995) Arch Biochem Biophys **319**: 333
[9] Hu S, Hager LP (1999) J Am Chem Soc **121**: 872
[10] Corbett MD, Chipko BR (1979) Biochem J **183**: 269
[11] Van Deurzen MPJ, Remkes IJ, van Rantwijk F, Sheldon RA (1997) J Mol Cat A: Chemical **117**: 329
[12] Van Deurzen MPJ, Seelbach K, van Rantwijik F, Kragl U, Sheldon RA (1997) Biocat Biotrans **15**: 1
[13] De Gioia L, Ghibaudi EM, Laurenti E, Salmona M, Ferrari RP (1996) J Bioinorg Chem **1**: 476
[14] Klebanoff SJ (1968) J Bacteriol **95**: 2131
[15] Mayfield MB, Kishi K, Alic M, Gold MH (1994) Appl Environ Microbiol **60**: 4303
[16] Ortiz de Montellano PR (1992) Annu Rev Pharmacol Toxicol **32**: 89
[17] Kobayashi S, Nakano M, Goto T, Kimura T, Schaap AP (1986) Biochem Biophys Res Commun **135**: 166
[18] Kobayashi S, Nakano M, Kimura T, Schaap PA (1987) Biochemistry **26**: 5019
[19] Casella L, Gullotti M, Ghezzi R, Poli S, Beringhelli T, Colonna S, Carrea G (1992) Biochemistry **31**: 9451
[20] Baciocchi E, Lanzalunga O, Malandrucco S (1996) J Am Chem Soc **118**: 8973
[21] Perez U, Dunford HB (1990) Biochem Biophys Acta **1038**: 98
[22] Perez U, Dunford, HB (1990) Biochemistry **29**: 2757
[23] Huwiler M, Jenzer H, Kohler H (1986) Eur J Biochem **158**: 609
[24] Tanaka M, Ishimori K, Mukai, M, Kitagawa T, Morishima I (1997) Biochemistry **36**: 9889
[25] Ozaki S, Ortiz de Montellano PRO (1994) J Am Chem Soc **116**: 4487
[26] Harris RZ, Newmyer SL, Ortiz de Montellano PR (1993) J Biol Chem **268**: 1637
[27] Ozaki S, Ortiz de Montellano PR (1995) J Am Chem Soc **117**: 7056
[28] Newmyer SL, Ortiz de Montellano PR (1995) J Biol Chem **270**: 19430
[29] Savenkova MI, Ortiz de Montellano PR (1998) Arch Biochem Biophys **351**: 286
[30] Savenkova MI, Ortiz de Montellano PR (1998) Biochemistry **37**: 10828
[31] Capeillère-Blandin C, Martin C, Gaggero N, Pasta P, Carrea G, Colonna S (1998) Biochem J **335**: 27
[32] Baunsgaard L, Dalbøge H, Houen G, Rasmussen EM, Welinder KG (1993) Eur J Biochem **213**: 605
[33] Bakkenist ARJ, Wever R, Vulsma T, Plat H, van Gelder BF (1978) Biochim Biophys Acta **524**: 45
[34] Schonbaum GR, Lo S (1972) J Biol Chem **247**: 3353
[35] Millis CD, Cai D, Stankovich MT, Tien M (1994) Biochemistry **28**: 8484
[36] Beers RF Jr and Sizer IW (1952) J Biol Chem **195**: 133

Received November 18, 1999. Accepted January 21, 2000

Synthesis of Optically Active α-Methyl β-Hydroperoxy Esters by Diastereoselective Singlet Oxygen Ene Reaction and Horseradish Peroxidase Catalyzed Kinetic Resolution

Waldemar Adam*, **Chantu R. Saha-Möller,** and **Oliver Weichold**

Institut für Organische Chemie der Universität Würzburg, D-97074 Würzburg, Germany

Summary. Optically active diastereomeric β-hydroperoxy esters **4** have been prepared by singlet oxygen ene reaction of β,γ-unsaturated esters **3** and subsequent horseradish peroxidase (HRP) catalyzed kinetic resolution of the ene product. The highest enantiomeric excess (up to 95%) has been obtained for the isopropyl ester *threo*-**4c**, which establishes that the size of the remote ester functionality exercises appreciable control in the enantioselectivity of the enzymatic kinetic resolution.

Keywords. γ,δ-Unsaturated β-hydroperoxy esters; Horseradish peroxidase; Kinetic resolution; *Schenck* ene reaction; Singlet oxygen.

Introduction

During the last two decades, metal-activated hydroperoxides have been widely used as oxidants in enantioselective synthesis under catalytic conditions [1]. An achiral peroxide, *e.g.* *t*BuOOH or cumyl hydroperoxide, is usually employed in combination with a chiral auxiliary for asymmetric induction. Only recently [2] have optically active hydroperoxides been applied for the direct enantioselective oxygen transfer without chiral auxiliaries, because few methods are known for the preparation of enantiomerically enriched or even pure hydroperoxides, most notably enzymatic resolution [3]. In this context we have developed a convenient, versatile, and quite general kinetic resolution of racemic hydroperoxides by horseradish peroxidase (HRP) in the presence of guajacol [4]. Substrates with varying size as well as structure and with diverse functionalities have been extensively examined by this method; however, the influence of remote groups of different size on the stereochemical consequences on the efficacy of the kinetic resolution has not yet been addressed. For this purpose, we have studied by HRP-catalyzed kinetic resolution of α-methyl β-hydroperoxy esters in which the size of the ester group (-CO_2R), remote from the chirality center, has been varied in the increasing order of methyl, ethyl, isopropyl, and isobutyl.

* Corresponding author

Results and Discussion

The photooxygenation of the *Z*-configured β,γ-unsaturated esters **3a–d** afforded the diastereomeric β-hydroperoxy esters **4a–d** in good yields by the singlet oxygen ene reaction (Scheme 1). Additionally, minor amounts of the regioisomeric acrylic acid derivatives **4a′–d′** were formed. The starting materials were prepared according to literature procedures. By starting from malonic acid and propionic aldehyde (*i*), (*E*)-2-pentenoic acid was formed [5] which was esterified in two ways (*ii*): the methyl, ethyl, and isopropyl esters were synthesized by means of the acid-catalyzed method of *Buchta* and *Burger* [6]; for the isobutyl derivative, to avoid possible rearrangement, the inverse method of *Johnstone* and *Rose* [7] was applied, *i.e.* alkylation of the carboxylate by isobutyl bromide. The deconjugative alkylation (*iii*) was performed in analogy to the literature procedure [8], in which the lithium enolate was allowed to react with methyl iodide; however, *HMPT* was replaced by the cyclic urea *DMPU* [9] without significant loss of yield.

The photooxygenation [10] (*iv*) of the *Z*-configured β,γ-unsaturated esters **3a–d** was performed in CCl_4 at $-20°C$ (Table 1) and afforded the diastereomeric

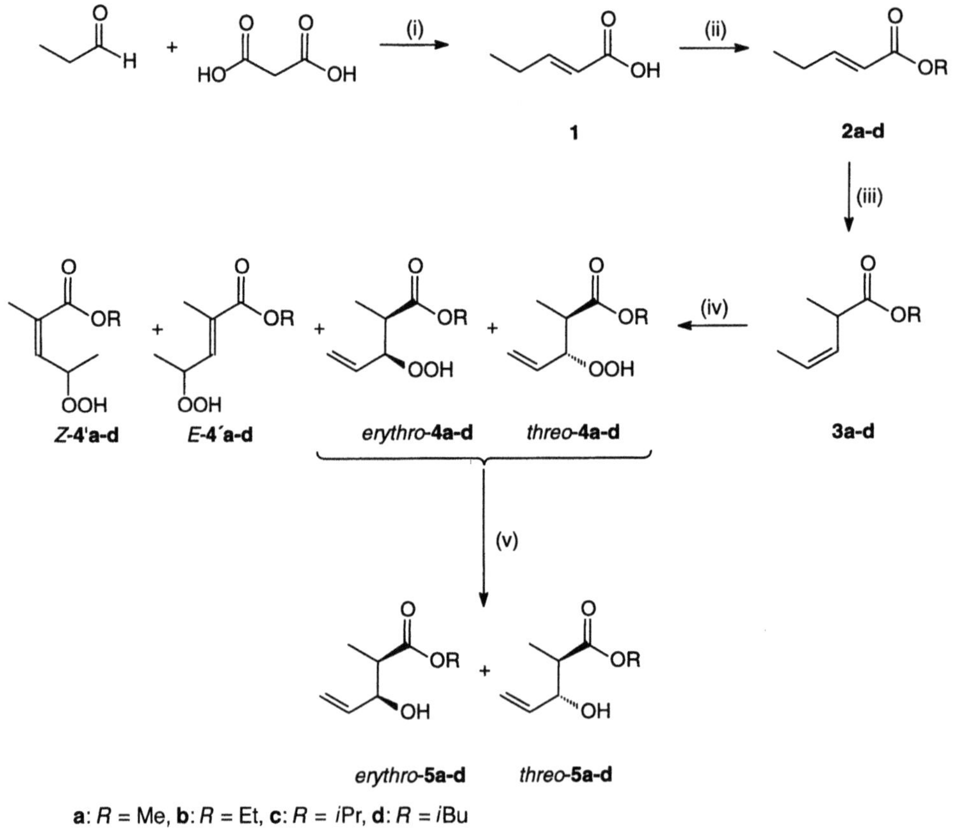

(*i*) Pyridine, 6d, 20°C, exclusion of light, 53% (Ref. [5]: 88%); (*ii*) *R* = Me, Et, *i*Pr: *R*OH, H_2SO_4, reflux [6]; *R* = *i*Bu: a) KOH, H_2O, 20°C, 2 h, b) *i*BuBr, *DMSO*, 20°C [7]; (*iii*) a) *LDA*/*DMPU*, *THF*, −78°C, 30 min, b) MeI, *THF*, −78°C, 45 min [8]; (*iv*) O_2, *TPP*, $h\nu$ (500 W), CCl_4, −20°C, 17 h [10]; (*v*) Ph_3P, Et_2O, 0°C, 30 min [11]

Scheme 1. Synthesis of β,γ-unsaturated esters, photooxygenation, and triphenylphosphine reduction

Table 1. Regio- and diastereoselectivities for the photooxygenation of the β, γ-unsaturated esters **3a–d**

				Selectivity[a]		
		Conversion[a]	Yield[b]	Regioselectivity	Diastereoselectivity[c]	
Entry	Substrate	[%]	[%]	**4**:**4′**	*erythro*:*threo*	*E*:*Z*
1	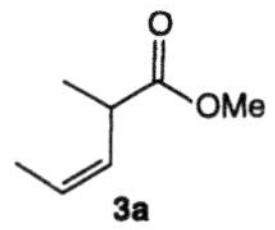 **3a**	92	65	90:10	77:23	34:66
2	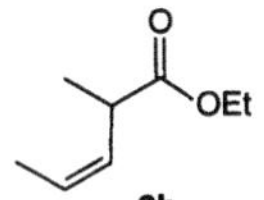 **3b**	>95	60	90:10	77:23	33:67
3	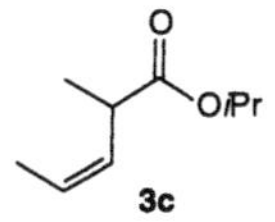 **3c**	>95	70	90:10	76:24	22:78
4	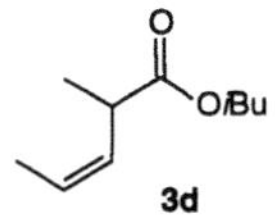 **3d**	>95	67	>95:5	80:20	–[d]

[a] Determined by ^{1}H NMR spectroscopy, error ±5%; [b] yield of isolated product **4** after silica gel chromatography; [c] relative configuration was assigned according to Ref. [10]. [d] *E/Z* isomers not detected

hydroperoxy esters **4a–d** in 60–70% yield, with the *erythro* isomers as the main products. Due to the 1,3-allylic strain in the *cis*-configured esters **3**, the conformer **A′** (Scheme 2) is less favoured, and **A** dominates. The incoming $^1O_2$ may attack from the two diastereomeric π faces. In exciplex **B′** (top attack), the electrostatic repulsion between the dioxygen molecule and the carbonyl group disfavours this attack and, therefore, the resulting *thero* hydroperoxy esters are formed as the minor isomers (20–23%); in exciplex **B** (bottom attack), such electrostatic interaction is avoided; consequently, the *erythro* isomers are formed as the main products (76–80%). Since in the exciplex **B** the ester group points away from the reaction center, the diastereomeric ratios are unaffected by the size of this group, as demonstrated by the fact that the *threo*/*erythro* ratio is approximately the same for the whole series of hydroperoxy esters **4a–d** (Table 1). Minor amounts (*ca.* 10%) of the **4′** regioisomers were also formed, which were separated from the main **4** regioisomers by flash chromatography on silica gel. All attempts to separate the diastereomers **4a–d** by any chromatographic method failed; therefore, the hydroperoxy esters were used as mixture for further transformations. For GC analysis, the hydroperoxy esters **4a–d** were reduced (Scheme 1, step *v*) by triphenylphosphine in ether at 0°C to afford the hydroxy esters **5a–d** in high yields (80–98%) [11].

The HRP-catalyzed kinetic resolution [4] of the hydroperoxy esters **4a–d** as *erythro*/*threo* diastereomeric mixtures (*ca.* 77:23, Table 1) was performed on a semi-preparative scale. The methyl, ethyl, and isopropyl esters were converted to

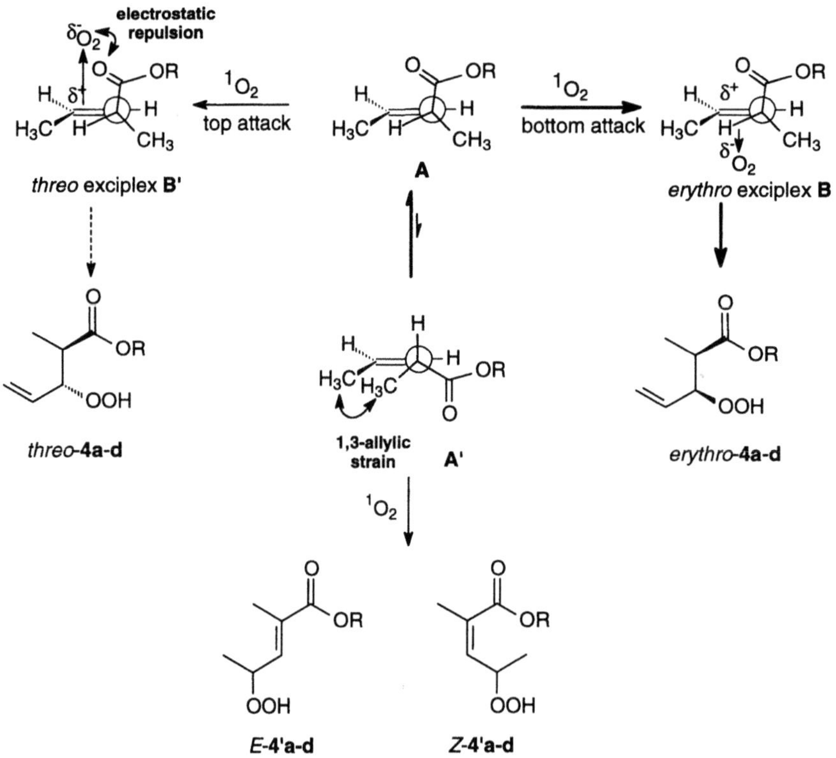

Scheme 2. Mechanism of the singlet oxygen ene reaction for the β,γ-unsaturated esters **3a–d**

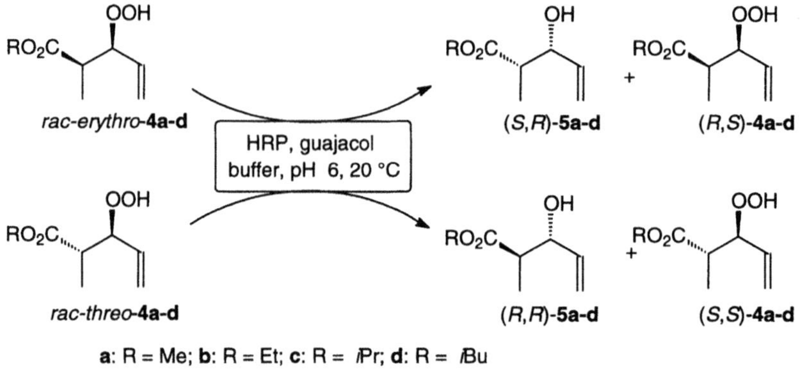

Scheme 3. HRP-catalyzed kinetic resolution of the racemic hydroperoxy esters **4a–d**; for the assignment of the absolute configuration, consult Scheme 4

the extent of *ca.* 50% within 4–6 h at ambient temperature (*ca.* 20°C) in a 0.1 *M* aqueous phosphate buffer ($pH = 6$) with 0.5 equiv. of guajacol and 1/10000 equiv. of enzyme (Scheme 3); in contrast, the isobutyl ester **4d** did not react under these conditions (Table 2). The conversions were determined by ^{1}H NMR spectroscopy on the crude product mixture and, for comparison, also calculated from the *ee* values of the hydroperoxides **4** and alcohols **5** (Table 2) according to

$$convn = 100 \times (ee_{\mathrm{ROOH}}/ee_{\mathrm{ROOH}} + ee_{\mathrm{ROH}}) \qquad (1)$$

Table 2. Enantioselectivities for the kinetic resolution of the hydroperoxy esters **4a–d**

				Conversion (%)		ee^{a}(%)		
Entry	Substrate		*t*/h	^{1}H NMR[b]	Eq. (1)[c]	**5**	**4**	*E*[d]
1	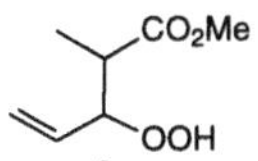	*erythro*	4	40	39	86 ± 1	53 ± 1	18
		threo			41	88 ± 1	60 ± 2	25
2[e]		*erythro*	4	24	23	82 ± 1	25 ± 1	14
	4a	*threo*			23	82 ± 1	24 ± 1	12
3	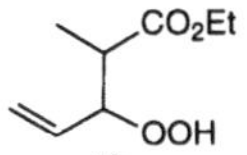	*erythro*	4	60	48	81 ± 1	76 ± 2	23
	4b	*threo*			–	–[f]	80 ± 1	34
4	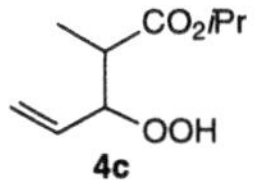	*erythro*	6	46	49	93 ± 1	88 ± 1	66
	4c	*threo*			49	>95	93 ± 1	>200
5	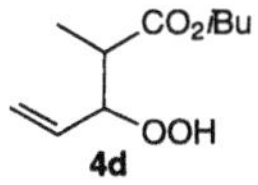	*erythro*	6	–[g]	–	–	–	
	4d	*threo*						

[a] The *ee* values were determined by chiral GC analysis, error limits refer to several determinations, absolute configuration was not determined; [b] yield determined by ^{1}H NMR spectroscopy on the crude product mixture, error *ca.* 5%; [c] conversion calculated according to Eq. (1), [d] values for **4** calculated according to $E = \ln((1 - convn) \times (1 - ee))/\ln((1 - convn) \times (1 + ee))$ [12]; [e] reaction performed in a 1:1 phosphate buffer (*pH* 6)/EtOH mixture; [f] the *threo* diastereomer was not detected; [g] no conversion

The *ee* values of the hydroxy esters **5**, which were separated from the hydroperoxy esters **4** by silica gel chromatography, were determined by GC analysis on a chiral adsorbent. The *ee* values of the hydroperoxy esters **4** were determined in the same way, but after triphenylphosphine reduction.

The size of the remote ester group does influence the enantioselectivity of the HRP-catalyzed resolution of the hydroperoxides **4** (Table 2), since for the methyl *vs.* isopropyl substrates the *ee* values of the alcohols and hydroperoxides rose from 86% (**5a**) and 53% (**4a**) to 93% (**5c**) and 88% (**4c**). However, in all cases no significant difference in *ee* values was observed for the *erythro* and *threo* diastereomers. Unfortunately, the large isobutyl group (**4d**) was not accepted by the enzyme. To enhance the solubility of the substrate **4a**, the pure phosphate buffer was replaced by a 1:1-mixture of buffer and EtOH; however, the conversion was significantly decreased (Table 2, entry 2). This may be ascribed to the favoured active conformation of the enzyme under the natural conditions of the aqueous buffered medium [13].

The determination of the absolute configuration by spectroscopic and/or chemical means was made difficult by the fact that the diastereomeric mixtures of the hydroperoxy (**4**) and hydroxy (**5**) esters could not be separated and, thus, the pure diastereomers were not accessible. For this reason, the configurations, given in Scheme 3 were tentatively assigned by analogy to the established configurations of optically active hydroperoxides obtained previously by HRP-catalyzed kinetic resolutions. The following empirical rule has been proposed [14]: The two α

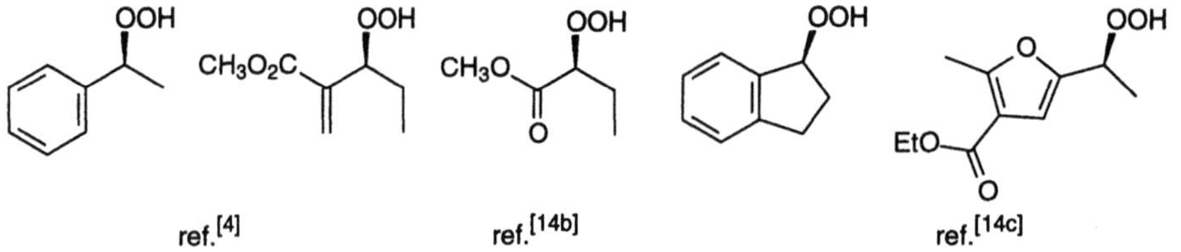

threo-**4a-d** *erythro*-**4a-d**

Scheme 4. Correlation of the absolute configuration of the hydroperoxy esters **4** with hydroperoxides of known configuration

carbon atoms of the chiral hydroperoxid are placed in the plane of the paper, with the small substituent on the right-hand and the larger one on the left-hand side (Scheme 4). The enantiomer with the hydroperoxy group above the paper plane is less likely reduced by the enzyme and, therefore, accumulates enantiomerically enriched in the HRP-catalyzed kinetic resolution. Application of this empirical rule leads to the absolute configurations of *threo*- and *etythro*-**4a–c** as depicted in Scheme 4, in which the vinyl group unquestionably qualifies as the smaller substituent compared to the ester-bearing group.

In conclusion, the present study clearly demonstrates that besides α substituents, also remote groups may influence the stereochemical course of the HRP-catalyzed reduction of chiral hydroperoxides. This affords attractive opportunities of controlling the enantioselectivity of such enzymatic transformations by attaching easily removable groups remote from the reaction center.

Experimental

THF was distilled from potassium metal, diisopropyl amine was refluxed over CaH_2 for one week and distilled subsequently. *DMPU* was dried according to literature [9]. For the photooxygenation, CCl_4 was distilled from P_2O_5 and the oxygen gas was dried over $CaCl_2$ and P_2O_5. Other commercially available reagents were used without further purification. The photooxygenations were performed at -20°C by irradiation with two 500 W sodium vapor lamps (Osram Vialox NAV-E) without filter. The horseradish peroxidase was obtained from Sigma (peroxidase type II, 1.5–2 U/mg).

Preparation of the starting materials

E-2-Pentenoic acid (**1**) was prepared from propionoic aldehyde and malonic acid in 53% yield [5]. The esters **2a–c** were synthesized according to literature in 73–80% yield. **2d** [7] and the *Z*-2-methylpentenoic esters **3a–d** [8] were obtained in yields of 91 and 46–70%, respectively. Photooxygenations were performed as reported [10] to afford the hydroperoxy esters **4a–d** in 60–70% yields as a 90:10-mixture of the regioisomers **4** and **4′** and a 77:23 mixture of the *erythro/threo* diastereomers. The reduction to the hydroxy esters **5a–d** was performed as described [11] in 80–98% yield. Compounds **1** [5], **2a** [6], **b** [5], **c** [15], **d** [15], **3a** [16], **b** [8], **c** [16], **d** [16], **4b** [10], **5a**[17], **b**

[18] are known. No satisfactory elemental analyses could be obtained for the labile hydroperoxides **4a** and **4d**; the peroxide content was therefore determined iodometrically.

Methyl erythro, threo-3-hydroperoxy-2-methyl-4-pentenoates (**4a**; $C_7H_{12}O_4$)
and *(E,Z)-4-Hydroperoxy-2-methyl-2-pentenoates* (**4a**′; $C_7H_{12}O_4$)

Peroxide content: >95%; *erythro*-**4a**: ^{1}H NMR ($CDCl_3$, δ, 250 MHz): 1.19 (d, $J_{6,2} = 7.2$ Hz, 3H, 6-H), 2.84 (qd, $J_{2,6} = 7.2$ Hz, $J_{2,3} = 6.5$ Hz, 1H, 2-H), 3.69 (s, 3H, 7-H), 4.58 (ddm, $J_{3,4} = 7.5$ Hz, $J_{3,2} = 6.5$ Hz, 1H, 3-H), 5.32 (m, 1H, 5-H_b), 5.38 (ddd, $J_{5a,4} = 17.4$ Hz, $J_{5a,5b} = J_{5a,3} = 0.9$ Hz, 1H, 5-H_a), 5.82 (ddd, $J_{4,5a} = 17.4$ Hz, $J_{4,5b} = 10.4$ Hz, $J_{4,3} = 7.5$ Hz, 1H, 4-H), 8.49 (s, 1H, OOH) ppm; ^{13}C NMR ($CDCl_3$, δ, 63 MHz): 12.2 (q, C-6), 42.3 (d, C-2), 52.0 (q, C-7), 87.3 (d, C-3), 120.7 (t, C-5), 133.4 (d, C-4), 174.2 (C-1) ppm; *threo*-**4a**: ^{1}H NMR ($CDCl_3$, δ, 250 MHz): 1.12 (d, $J_{6,2} = 7.2$ Hz, 3H, 6-H), 2.84 (qd, $J_{2,6} = 7.2$ Hz, $J_{2,3} = 6.5$ Hz, 1H, 2-H), 3.69 (s, 3H, 7-H), 4.50 (ddm, $J_{3,4} = 7.9$ Hz, $J_{3,2} = 6.5$ Hz, 1H, 3-H), 5.32 (m, 1H, 5-H_b), 5.38 (ddd, $J_{5a,4} = 17.4$ Hz, $J_{5a,5b} = J_{5a,3} = 0.9$ Hz, 1H, 5-H_a), 5.77 (ddd, $J_{4,5a} = 17.4$ Hz, $J_{4,5b} = 10.4$ Hz, $J_{4,3} = 7.9$ Hz, 1H, 4-H), 8.49 (s, 1H, OOH) ppm; ^{13}C NMR ($CDCl_3$, δ, 63 MHz): 13.2 (q, C-6), 41.9 (d, C-2), 52.0 (q, C-7), 88.2 (d, C-3), 121.4 (t, C-5), 133.4 (d, C-4), 174.2 (C-1) ppm; the signals for the regioisomers (*E*)-**4a**′ and (*Z*)-**4a**′ could not be assigned due to severe overlap in the NMR spectrum of the crude product; IR (neat): $\nu = 3640$-3020 (OOH), 2940, 2910, 1700 (C=O), 1435, 1410, 1360, 1330, 1240, 1190, 1040, 980, 920, 860 cm^{-1}.

Isopropyl erythro, threo-3-hydroperoxy-2-methyl-4-pentenoates (**4c**; $C_9H_{16}O_4$)
and (E,Z)-4-hydroperoxy-2-methyl-2-pentenoates (**4c**′; $C_9H_{16}O_4$)

Calcd.: C 57.43, H 8.57; found: C 57.16, H 8.80; *erythro*-**4c**: ^{1}H NMR ($CDCl_3$, δ, 250 MHz): 1.23 (d, $J_{6,2} = 7.2$ Hz, 3H, 6-H), 1.23 (d, $J_{8,7} = 6.3$ Hz, 6H, 8-H), 2.78 (qd, $J_{2,6} = 7.2$ Hz, $J_{2,3} = 6.4$ Hz, 1H, 2-H), 4.57 (ddm, $J_{3,4} = 7.4$ Hz, $J_{3,2} = 6.4$ Hz, 1H, 3-H), 5.02 (sept, $J_{7,8} = 6.3$ Hz, 1H, 7-H), 5.33 (dm, $J_{5b,4} = 10.2$ Hz, 1H, 5-H_b), 5.39 (dm, $J_{5a,4} = 18.9$ Hz, 1H, 5-H_a), 5.83 (ddd, $J_{4,5a} = 18.9$ Hz, $J_{4,5b} = 10.2$ Hz, $J_{4,3} = 7.4$ Hz, 1H, 4-H), 8.43 (s, 1H, 3-OOH) ppm; ^{13}C NMR ($CDCl_3$, δ, 63 MHz): 12.3 (q, C-6), 21.7 (2q, C-8), 42.5 (d, C-2), 68.2 (d, C-7), 87.3 (d, C-3), 120.5 (t, C-5), 133.6 (d, C-4), 173.3 (s, C-1) ppm; *threo*-**4c**: ^{1}H NMR ($CDCl_3$, δ, 250 MHz): 1.19 (d, $J_{6,2} = 7.2$ Hz, 3H, 6-H), 1.23 (d, $J_{8,7} = 6.3$ Hz, 6H, 8-H), 2.78 (qd, $J_{2,6} = 7.2$ Hz, $J_{2,3} = 6.4$ Hz, 1H, 2-H), 4.48 (ddm, $J_{3,4} = 8.0$ Hz, $J_{3,2} = 6.4$ Hz, 1H, 3-H), 5.02 (sept, $J_{7,8} = 6.3$ Hz, 1H, 7-H), 5.33 (dm, $J_{5b,4} = 10.2$ Hz, 1H, 5-H_b), 5.39 (dm, $J_{5a,4} = 18.9$ Hz, 1H, 5-H_a), 5.83 (ddd, $J_{4,5a} = 18.9$ Hz, $J_{4,5b} = 10.2$ Hz, $J_{4,3} = 8.0$ Hz, 1H, 4-H), 8.43 (s, 1H, OOH) ppm; ^{13}C NMR ($CDCl_3$, δ, 63 MHz): 12.3 (q, C-6), 21.7 (2q, C-8), 42.5 (d, C-2), 68.2 (d, C-7), 88.2 (d, C-3), 121.2 (t, C-5), 133.6 (d, C-4), 173.3 (s, C-1) ppm; the signals for the regioisomers (*E*)-**4c**′ and (*Z*)-**4c**′ could not be assigned due to severe overlap in the NMR spectrum of the crude product; IR (neat): $\nu = 3680$-3100 (OOH), 2960, 2920, 2860, 1710 (C=O), 1690, 1440, 1365, 1100, 985, 820 cm^{-1}.

Isobutyl erythro, threo-3-hydroperoxy-2-methyl-4-pentenoates (**4d**; $C_{10}H_{18}O_4$)

Peroxide content: 91%; *erythro*-**4d**: ^{1}H NMR ($CDCl_3$, δ, 250 MHz): 0.93 (d, $J_{9,8} = 6.7$ Hz, 6H, 9-H), 1.21 (d, $J_{6,2} = 7.0$ Hz, 3H, 6-H), 1.94 (m, 1H, 8-H), 2.84 (qd, $J_{2,6} = 7.0$ Hz, $J_{2,3} = 6.2$ Hz, 1H, 2-H), 3.90 (d, $J_{7,8} = 6.7$ Hz, 2H, 7-H), 4.58 (ddm, $J_{3,4} = 7.4$ Hz, $J_{3,2} = 6.2$ Hz, 1H, 3-H), 5.33 (ddd, $J_{5b,4} = 4.3$ Hz, $J_{5b,5a} = 1.5$ Hz, $J_{5b,3} = 0.9$ Hz, 1H, 5-H_b), 5.39 (ddd, $J_{5a,4} = 11.3$ Hz, $J_{5a,5b} = 1.5$ Hz, $J_{5a,3} = 0.9$ Hz, 1H, 5-H_a), 5.84 (ddd, $J_{4,5a} = 11.3$ Hz, $J_{4,3} = 7.4$ Hz, $J_{4,5b} = 4.3$ Hz, 1H, 4-H), 8.46 (s, 1H, OOH) ppm; ^{13}C NMR ($CDCl_3$, δ, 63 MHz): 12.1 (q, C-6), 19.0 (2q, C-9), 27.7 (d, C-8), 42.1 (d, C-2), 71.0 (t, C-8), 87.2 (d, C-3), 120.5 (t, C-5), 133.7 (d, C-4), 173.9 (s, C-9) ppm; *threo*-**4d**: ^{1}H NMR ($CDCl_3$, δ, 250 MHz): 0.94 (d, $J_{9,8} = 6.7$ Hz, 6H, 9-H), 1.12 (d, $J_{6,2} = 7.0$ Hz, 3H, 6-H), 1.94 (m, 1H, 8-H), 2.75 (qd, $J_{2,6} = 7.0$ Hz, $J_{2,3} = 6.2$ Hz, 1H, 2-H), 3.87 (d, $J_{7,8} = 6.7$ Hz, 2H, 7-H), 4.52 (ddm, $J_{3,4} = 8.5$ Hz, $J_{3,2} = 6.2$ Hz, 1H, 3-H), 5.33 (ddd, $J_{5b,4} = 4.3$ Hz, $J_{5b,5a} = 1.5$ Hz, $J_{5b,3} = 0.9$ Hz, 1H,

5-H_b), 5.39 (ddd, $J_{5a,4} = 11.3$ Hz, $J_{5a,5b} = 1.5$ Hz, $J_{5a,3} = 0.9$ Hz, 1H, 5-H_a), 5.80 (ddd, $J_{4,5a} = 11.3$ Hz, $J_{4,3} = 8.5$ Hz, $J_{4,5b} = 4.3$ Hz, 1H, 4-H), 8.46 (s, 1H, OOH) ppm; ^{13}C NMR ($CDCl_3$, δ, 63 MHz): 13.2 (q, C-6), 19.0 (2q, C-9), 27.7 (d, C-8), 42.5 (d, C-2), 71.0 (t, C-7), 88.2 (d, C-3), 121.2 (t, C-5), 133.7 (d, C-4), 174.3 (s, C-9) ppm; IR (neat): ν = 3620-3100 (OOH), 2920, 2820, 1700 (C=O), 1620, 1440, 1410, 1230, 1175, 1030, 975, 910, 850 cm^{-1}.

Isopropyl erythro, threo-3-hydroxy-2-methyl-4-pentenoates (**5c**; $C_9H_{16}O_3$)

Calcd.: C 62.77, H 9.36; found: C 62.88, H 9.27; *erythro*-**5c**: ^{1}H NMR ($CDCl_3$, δ, 250 MHz): 1.15 (d, $J_{6,2} = 7.3$ Hz, 3H, 6-H), 1.23 (d, $J_{8,7} = 6.3$ Hz, 6H, 8-H), 2.58 (qd, $J_{2,6} = 7.3$ Hz, $J_{2,3} = 4.0$ Hz, 1H, 2-H), 4.38 (m, 1H, 3-H), 5.04 (sept, $J_{7,8} = 6.3$ Hz, 1H, 7-H), 5.19 (ddd, $J_{5b,4} = 10.5$ Hz, $J_{5b,5a} = J_{5b,3} = 1.6$ Hz, 1H, 5-H_b), 5.32 (ddd, $J_{5a,4} = 17.2$ Hz, $J_{5a,5b} = J_{5a,3} = 1.6$ Hz, 1H, 5-H_a), 5.82 (ddd, $J_{4,5a} = 17.2$ Hz, $J_{4,5b} = 10.5$ Hz, $J_{4,3} = 5.5$ Hz, 1H, 4-H) ppm, OH not detected; ^{13}C NMR ($CDCl_3$, δ, 63 MHz): 11.2 (q, C-6), 21.7 (2q, C-8), 44.6 (d, C-2), 68.1 (d, C-7), 73.0 (d, C-3), 116.2 (t, C-5), 137.3 (d, C-4), 174.9 (s, C-1) ppm; *threo*-**5c**: ^{1}H NMR ($CDCl_3$, δ, 250 MHz): 1.17 (d, $J_{6,2} = 7.3$ Hz, 3H, 6-H), 1.40 (d, $J_{8,7} = 6.7$ Hz, 6H, 8-H), 2.58 (qd, $J_{2,6} = 7.3$ Hz, $J_{2,3} = 4.0$ Hz, 1H, 2-H), 4.38 (m, 1H, 3-H), 5.04 (sept, $J_{7,8} = 6.7$ Hz, 1H, 7-H), 5.19 (ddd, $J_{5b,4} = 10.5$ Hz, $J_{5b,5a} = J_{5b,3} = 1.6$ Hz, 1H, 5-H_b), 5.30 (ddd, $J_{5a,4} = 17.1$ Hz, $J_{5a,5b} = J_{5a,3} = 1.6$ Hz, 1H, 5-H_a), 5.83 (ddd, $J_{4,5a} = 17.1$ Hz, $J_{4,5b} = 10.5$ Hz, $J_{4,3} = 5.5$ Hz, 1H, 4-H) ppm, OH not detected; ^{13}C NMR ($CDCl_3$, δ, 63 MHz): 14.0 (q, C-6), 30.9 (2q, C-8), 45.2 (d, C-2), 68.1 (d, C-7), 74.7 (d, C-3), 116.7 (t, C-5), 138.2 (d, C-4), 174.9 (s, C-1) ppm; IR (neat): ν = 3620-3100 (OH), 2960, 2920, 2860, 1710 (C=O), 1690 (C=C), 1440, 1365, 1250, 1185, 1100, 920 cm^{-1}.

Isobutyl erythro,threo-3-hydroxy-2-methyl-4-pentenoates (**5d**; $C_{10}H_{18}O_3$)

Calcd.: C 64.49, H 9.74; found: C 64.22, H 9.56; *erythro*-**5d**: ^{1}H NMR ($CDCl_3$, δ, 200 MHz): 0.86 (d, $J_{9,8} = 6.7$ Hz, 6H, 9-H), 1.11 (d, $J_{6,2} = 7.2$ Hz, 3H, 6-H), 1.87 (septm, $J_{8,9} = 6.7$ Hz, 1H, 8-H), 2.55 (qd, $J_{2,6} = 7.2$ Hz, $J_{2,3} = 4.6$ Hz, 1H, 2-H), 3.81 (d, $J_{7,8} = 6.6$ Hz, 2H, 7-H), 4.32 (m, 1H, 3-H), 5.10 (ddd, $J_{5b,4} = 10.5$ Hz, $J_{5b,5a} = J_{5b,3} = 1.5$ Hz, 1H, 5-H_b), 5.23 (ddd, $J_{5a,4} = 17.2$ Hz, $J_{5a,5b} = J_{5a,3} = 1.5$ Hz, 1H, 5-H_a), 5.77 (ddd, $J_{4,5a} = 17.2$ Hz, $J_{4,5b} = 10.5$ Hz, $J_{4,3} = 5.6$ Hz, 1H, 4-H), ppm, OH not detected; ^{13}C NMR ($CDCl_3$, δ, 50 MHz): 11.3 (q, C-6), 18.9 (2q, C-9), 27.5 (d, C-8), 44.7 (d, C-2), 70.5 (t, C-7), 73.0 (d, C-3), 115.9 (t, C-5), 137.5 (d, C-4), 175.1 (s, C-1) ppm; *threo*-**5d**: ^{1}H NMR ($CDCl_3$, δ, 200 MHz): 0.86 (d, $J_{9,8} = 6.7$ Hz, 6H, 9-H), 1.10 (d, $J_{6,2} = 7.2$ Hz, 3H, 6-H), 1.87 (septm, $J_{8,9} = 6.7$ Hz, 1H, 8-H), 2.55 (qd, $J_{2,6} = 7.2$ Hz, $J_{2,3} = 4.6$ Hz, 1H, 2-H), 3.82 (d, $J_{7,8} = 6.6$ Hz, 2H, 7-H), 4.32 (m, 1H, 3-H), 5.10 (ddd, $J_{5b,4} = 10.5$ Hz, $J_{5b,5a} = J_{5b,3} = 1.5$ Hz, 1H, 5-H_b), 5.23 (ddd, $J_{5a,4} = 17.2$ Hz, $J_{5a,5b} = J_{5a,3} = 1.5$ Hz, 1H, 5-H_a), 5.77 (ddd, $J_{4,5a} = 17.2$ Hz, $J_{4,5b} = 10.5$ Hz, $J_{4,3} = 5.6$ Hz, 1H, 4-H), ppm, OH not detected; ^{13}C NMR ($CDCl_3$, δ, 50 MHz): 13.7 (q, C-6), 18.9 (2q, C-9), 27.5 (d, C-8), 45.2 (d, C-2), 70.6 (t, C-6), 74.5 (d, C-3), 116.7 (t, C-5), 138.0 (d, C-4), 175.3 (s, C-1) ppm; IR (neat): ν = 3600-3120, 2940, 2850, 1710 (C=O), 1690 (C=C), 1450, 1360, 1330, 1250, 1180, 1040, 980, 910 cm^{-1}.

General procedure for HRP-catalyzed kinetic resolution on a semi-preparative scale

Guajacol (0.05–0.30 mmol, 0.5 equiv.) and HRP ($1–3 \times 10^{-5}$ mmol, 1/10000 equiv.) were dissolved in 2 cm^3 of 0.1 *M* phosphate buffer (*pH* = 6), and subsequently 0.1–0.6 mmol of the particular hydroperoxy esters **4** (as diastereomeric mixture) were added. The reaction was stirred at *ca.* 20°C until about 50% conversion was achieved (*ca.* 4–6 h) and then extracted with CH_2Cl_2 (5×10 cm^3). The combined organic phases were dried over $MgSO_4$, and the solvent was evaporated (0°C, 12 torr). The products were separated by flash chromatography (40 g silica gel (0.063–0.200 mm), petroleum ether (30–50): ethyl ether = 9:1 as eluent) to afford the enantiomerically enriched hydroperoxy esters

4a–d and the corresponding hydroxy esters **5a–d**. The enantiomeric excess was determined by GC analysis on a Cyclodex-B column (30 m, 0,25 mm ID, H_2 gas). Conditions **5a**: 60°C (5 min), 5°C/min, 90°C; **5b**: 60°C (1 min), 5°C/min, 90°C; **5c,d**: 60°C (1 min), 5°C/min, 80°C.

Acknowledgements

The authors thank the *Bayerische Forschungsstiftung* (Bayerischer Forschungsverbund Katalyse-FORKAT), the *Deutsche Forschungsgemeinschaft* (SFB 347, "Selektive Reaktionen Metall-aktivierter Moleküle"), and the *Fonds der Chemischen Industrie* for generous support.

References

[1] a) Johnson RA, Sharpless KB (1993) In: Ojima I (ed) Catalytic Asymmetric Synthesis, VCH, Weinheim, p 103; b) Murahashi SI, Komiya N (1998) Catal Today **41**: 339

[2] a) Adam W, Korb MN, Roschmann KJ, Saha-Möller CR (1998) J Org Chem **63**: 3423; b) Hamann HJ, Höft E, Chmielewski M, Maciejewski S (1993) Chirality **5**: 338; c) Hamann HJ, Höft E, Mostowicz D, Mishnev A, Urbańczyk-Lipkowska Z, Chmielewski M (1997) Tetrahedron **53**: 185; d) Shum WP, Saxton RJ, Zajacek JG (1997) US Patent 5, 663, 384

[3] a) Baba N, Mimura M, Hiratake J, Uchida K, Oda J (1988) Agric Biol Chem **52**: 2685; b) Fu H, Kondo H, Ichikawa Y, Look GC, Wong CH (1992) J Org Chem **57**: 7265

[4] Adam W, Lazarus M, Saha-Möller CR, Weichold O, Hoch U, Häring D, Schreier P (1999) In: Faber K (ed) Advances in Biochemical Engineering/Biotechnology, vol 63. Springer, Heidelberg, p 73

[5] von Auwers K (1923) Chem Ber 46

[6] Buchta E, Burger K (1952) Justus Liebigs Ann Chem **576**: 156

[7] Johnstone RAW, Rose ME (1979) Pure Appl Chem **35**: 2169

[8] Kende AS, Toder BH (1982) J Org Chem **47**: 163

[9] Mukhopadhyay T, Seebach D (1982) Helv Chim Acta **65**: 385

[10] Adam W, Brünker HG, Sampath Kumar A, Peters EM, Peters K, Schneider U, von Schnering HG (1996) J Am Chem Soc **118**: 1899

[11] Adam W, Nestler B (1992) J Am Chem Soc **114**: 6549

[12] Chen CS, Fujimoto Y, Girdaukas G, Sih CJ (1982) J Am Chem Soc **104**: 7294

[13] Kazandjian RZ, Klibanov AM (1985) J Am Chem Soc **107**: 5448

[14] a) Adam W, Hoch U, Lazarus M, Saha-Möller CR, Schreier P (1995) J Am Chem Soc **117**: 11898; b) Hoch U, Adam W, Fell RT, Saha-Möller CR, Schreier P (1997) J Mol Cat A: Chemical **117**: 321; c) Adam W, Humpf HU, Korb M, Schreier P (1997) Tetrahedron Asymm **8**: 3555

[15] Schjanberg P (1937) Z Phys Chem A **179**: 32

[16] Tseng CY, Hall JB, Vock MH, Vinals JF, Shuster EJ, Hruza DE (1976) GE Patent 2530227

[17] Dodd DS, Oehlschlager AC, Georgopapadakou NH, Polak AM, Hartman PG (1992) J Org Chem **57**: 7226

[18] Smith AB, Levenberg PA (1981) Synthesis 567

Received December 20, 1999. Accepted (revised) February 11, 2000

GPSR Compliance

The European Union's (EU) General Product Safety Regulation (GPSR) is a set of rules that requires consumer products to be safe and our obligations to ensure this.

If you have any concerns about our products, you can contact us on ProductSafety@springernature.com

In case Publisher is established outside the EU, the EU authorized representative is:

Springer Nature Customer Service Center GmbH
Europaplatz 3
69115 Heidelberg, Germany

Batch number: 10731490

Printed by Printforce, the Netherlands